Lecture Notes
in Business Information Processing   **484**

Series Editors

Wil van der Aalst, *RWTH Aachen University, Aachen, Germany*
Sudha Ram, *University of Arizona, Tucson, AZ, USA*
Michael Rosemann, *Queensland University of Technology, Brisbane, QLD, Australia*
Clemens Szyperski, *Microsoft Research, Redmond, WA, USA*
Giancarlo Guizzardi, *University of Twente, Enschede, The Netherlands*

T0172224

LNBIP reports state-of-the-art results in areas related to business information systems and industrial application software development – timely, at a high level, and in both printed and electronic form.

The type of material published includes

- Proceedings (published in time for the respective event)
- Postproceedings (consisting of thoroughly revised and/or extended final papers)
- Other edited monographs (such as, for example, project reports or invited volumes)
- Tutorials (coherently integrated collections of lectures given at advanced courses, seminars, schools, etc.)
- Award-winning or exceptional theses

LNBIP is abstracted/indexed in DBLP, EI and Scopus. LNBIP volumes are also submitted for the inclusion in ISI Proceedings.

Rachid El Ayachi · Mohamed Fakir ·
Mohamed Baslam
Editors

Business Intelligence

8th International Conference, CBI 2023
Istanbul, Turkey, July 19–21, 2023
Proceedings

Editors
Rachid El Ayachi 🆔
Sultan Moulay Slimane University
Beni-Mellal, Morocco

Mohamed Fakir 🆔
Sultan Moulay Slimane University
Beni-Mellal, Morocco

Mohamed Baslam 🆔
Sultan Moulay Slimane University
Beni-Mellal, Morocco

ISSN 1865-1348   ISSN 1865-1356 (electronic)
Lecture Notes in Business Information Processing
ISBN 978-3-031-37871-3   ISBN 978-3-031-37872-0 (eBook)
https://doi.org/10.1007/978-3-031-37872-0

This Springer imprint is published by the registered company Springer Nature Switzerland AG
The registered company address is: Gewerbestrasse 11, 6330 Cham, Switzerland

Preface

After the success of the 7th edition, CBI 2022, this volume contains the proceedings of the 8th International Conference on Business Intelligence, CBI 2023, held during July 19–21, 2023, in Istanbul, Turkey.

This event was organized by the Faculty of Sciences and Technology (FST), Laboratory of Information Processing and Decision Support (TIAD), Sultan Moulay Slimane University, and by the Moroccan Association of Business Intelligence (AMID).

CBI has become the leading scientific forum for dissemination of cutting-edge research results in the area of Business Intelligence, which is one of the most vibrant areas of interest in modern society. It is a great occasion for researchers, engineers and managers from academia and industry to share their recent research results and to present and discuss their contributions.

This volume collects the papers accepted after a rigorous evaluation realized by the Program Committee, composed of 90 international experts in various fields related to business intelligence and decision support. In total, 50 submissions were received. The type of peer review used was Double Blind. Three reviewers reviewed each paper; however, we assigned two papers to each reviewer. The Program Committee decided to accept 15 regular papers, yielding an acceptance rate of 30%. The contributions are organized in two topical sections: "Optimization and Decision Support" and "Artificial Intelligence and Business Intelligence".

As chairs of the CBI 2023 program and editors of these proceedings, we would like to thank the President of Sultan Moulay Slimane University and the Dean of the Faculty of Science and Technology for their support of the conference. Furthermore, we would like to express our best gratitude to the keynote and invited speakers for their invaluable contribution and knowledge shared during the conference. In addition, we would like to thank all authors for their contribution as well as the Program Committee members for their constructive comments and suggestions for improving the quality documents presented. Finally, our special thanks go to the Organizing Committee for their invaluable hard work in making this edition of CBI a success.

We cordially invite you to visit the CBI website at https://www.cbibm.com/, enjoy reading this volume of proceedings, and join us in future editions of CBI.

July 2023

Rachid El Ayachi
Mohamed Fakir
Mohamed Baslam

Organization

General Chair

Rachid El Ayachi   Sultan Moulay Slimane University, Morocco

Program Chairs

Mohamed Baslam   Sultan Moulay Slimane University, Morocco
Mohamed Fakir   Sultan Moulay Slimane University, Morocco

Steering Committee

Driss Ait Omar	Sultan Moulay Slimane University, Morocco
Zafer Aslan	Istanbul Aydin University, Turkey
Ebad Banissi	London South Bank University, UK
Mohamed Biniz	Sultan Moulay Slimane University, Morocco
Charki Daoui	Sultan Moulay Slimane University, Morocco
A. Manuel De Oliveira Duarte	University of Aveiro, Portugal
Youssef El Mourabit	Sultan Moulay Slimane University, Morocco
Hammou Fadili	National Conservatory of Arts and Crafts, France
Hamid Garmani	Sultan Moulay Slimane University, Morocco
Kürşad Melih Güleren	Istanbul Aydin University, Turkey
Ali Güneş	Istanbul Aydin University, Turkey
Majed Haddad	University of Avignon, France
Suliman Hawamdeh	University of North Texas, USA
Celal Nazım İrem	Istanbul Aydin University, Turkey
Moustafa Jourhmane	Sultan Moulay Slimane University, Morocco
Brahim Minaoui	Sultan Moulay Slimane University, Morocco
Ali Okatan	Istanbul Aydin University, Turkey
Mohamed Ouhda	Sultan Moulay Slimane University, Morocco
Essaid Sabir	Hassan II University of Casablanca, Morocco
Abderrahim Salhi	Sultan Moulay Slimane University, Morocco
Muhammad Sarfraz	Kuwait University, Kuwait
Semih Yön	Istanbul Aydin University, Turkey
Amani Yusuf	İstanbul Sabahattin Zaim Üniversitesi, Turkey
Hicham Zougagh	Sultan Moulay Slimane University, Morocco

Program Committee

Abd Samad Hasan Basari	Universiti Teknikal Malaysia Melaka, Malaysia
Abdelali Elmoufidi	Sultan Moulay Slimane University, Morocco
Abdel-Badeeh Salem	University of Ain Shams, Egypt
Abdelhak Mahmoudi	Mohammed V University, Morocco
Abderrazak Farchane	Sultan Moulay Slimane University, Morocco
Akhil Kumar	Pennsylvania State University, USA
Akhilesh Bajaj	University of Tulsa, USA
Hamid Aksasse	FSTE, Morocco
Alberto Cano	Virginia Commonwealth University, USA
Alda Kika	University of Tirana, Albania
Ali Ouni	ETS Montreal, University of Quebec, Canada
Ankit Rajpal	University of Delhi, India
Ansuman Banerjee	Indian Statistical Institute, India
Antonio Lucadamo	University of Sannio, Italy
Athman Bouguettaya	University of Sydney, Australia
Aytuğ Onan	Izmir Katip Celebi University, Turkey
Bernhard Bauer	University of Augsburg, Germany
Bilge Karaçalı	İzmir Institute of Technology, Turkey
Bin Cao	Zhejiang University of Technology, China
Blerim Rexha	University of Pristina, Kosovo
Mohammed Boutalline	Sultan Moulay Slimane University, Morocco
Cemal Hanilçi	University of Bursa Technical, Turkey
Chaman Verma	Eötvös Loránd University, Hungary
Chaochao Chen	Zhejiang University of Technology, China
Dahmouni Abdellatif	FSDM, Morocco
Dilian Gurov	KTH Royal Institute of Technology, Sweden
Dirk Pattinson	Australian National University, Australia
Dmitry Chistikov	University of Warwick, UK
Dunwei Gong	University of Mining and Technology, China
Elinda Kajo Meçe	Polytechnic University of Tirana, Albania
Enayat Rajabi	University of Cape Breton, Canada
Esra Meltem Koç	Izmir Katip Celebi University, Turkey
Farokh Bastani	University of Texas at Dallas, USA
Farshad Firouzi	Duke University, USA
Francesco Regazzoni	Polytechnic University of Milan, Italy
Gabriel Bekdaş	University of İstanbul, Turkey
Gandhi Hernandez	University Carlos III of Madrid, Spain
Giancarlo Mauri	University of Milano-Bicocca, Italy
Giner Alor Hernandez	Technological Institute of Orizaba, Mexico
Meryeme Hadni	FSDM, Morocco

Hammou Fadili	National Conservatory of Arts and Crafts, France
Harald Kitzmann	University of Tartu, Kazakhstan
Hassan Silkan	Chouaib Doukkali University, Morocco
Hicham Mouncif	Sultan Moulay Slimane University, Morocco
Himadri Singh Raghav	National University of Singapore, Singapore
Khalid Housni	Ibn Tofail University, Morocco
Huaming Chen	University of Sydney, Australia
İbrahim Pirim	Izmir Katip Çelebi University, Turkey
Ismael Bouassida Rodriguez	University of Sfax, Tunisia
Jamal Hussain	University of Mizoram, India
Javier Berrocal	University of Extremadura, Spain
Jilali Antari	Ibn Zohr University, Morocco
Jochen Meyer	OFFIS - Institute for Information Technology, Germany
Jorge Sá Silva	University of Coimbra, Portugal
Juan A. Gómez-Pulido	University of Extremadura, Spain
Kenneth Fletcher	University of Massachusetts Boston, USA
Abdelmajid Khelil	University of Applied Sciences Landshut, Germany
Kyrre Glette	University of Oslo, Norway
Leila Ismail	United Arab Emirates University, United Arab Emirates
Matthew Hague	Royal Holloway, University of London, UK
M'hamed Outanoute	Sultan Moulay Slimane University, Morocco
Mohamed Erritali	Sultan Moulay Slimane University, Morocco
Mohammed Elamrani	Sultan Moulay Slimane University, Morocco
Mohd Ibrahim Shapiai Razak	Universiti Teknologi Malaysia, Malaysia
Müştak Erhan Yalçın	Istanbul Technical University, Turkey
Naoki Kobayashi	University of Tokyo, Japan
Nikitas N. Karanikolas	University of West Attica, Greece
Noureddine Falih	Sultan Moulay Slimane University, Morocco
Hicham Ouchitachen	Sultan Moulay Slimane University, Morocco
Pablo Barcelo	Universidad Católica de Chile, Chile
Pawel Sobocinski	Tallinn University of Technology, Estonia
Peyman Mahouti	İstanbul University-Cerrahpaşa, Turkey
Pinar Öztürk	Norwegian University of Science and Technology, Norway
Poonam Yadav	University of York, UK
Rami Bahsoon	University of Birmingham, UK
Sadok Ben Yahia	Tallinn University of Technology, Estonia
Said Safi	Sultan Moulay Slimane University, Morocco
Salwa Belaqziz	Ibn Zohr University, Morocco

Contents

Artificial Intelligence and Business Intelligence

A Business Intelligence System for Governing Risks in SMEs

Abdelaziz Darwiesh[1]([✉]) [iD], Mohamed Elhoseny[2] [iD], Reem Atassi[3] [iD],
and A. H. El-Baz[4] [iD]

[1] Faculty of Science, Damietta University, New Damietta 34517, Egypt
adarwiesh@isods.org
[2] Faculty of Computing and Informatics, University of Sharjah, Sharjah 26666, UAE
melhoseny@ieee.org
[3] Faculty of Computer Information System, Higher Colleges of Technology, Dubai 00000, UAE
[4] Faculty of Computers and Artificial Intelligence, Damietta University, New Damietta 34517,
Egypt

Abstract. This article suggests a new business intelligence system for risk management in small and medium-sized enterprises (SMEs). It depends on social media perceptions. It uses a lexicon approach to analyze, identify and assess risks. It will help SMEs managers in making best decisions. Also, a mathematical formulation and explicit formulas for the proposed system are provided. In addition, Book Nook enterprise is studied as a use case to examine the potential risks through the tweets of customers. Further, various performance indicators are calculated to validate the effectiveness of the suggested model.

Keywords: Business Intelligence · Risk Management · Social Media · SMEs

1 Introduction

SMEs represent a significant part in the economies of developed nations. They are highly exposed to problems in the business ecosystem that show up constantly over time in the quantitative behavior of this industry. In times of risk and uncertainty, managers are increasingly expected to make crucial decisions that will preserve profitability and financial stability [1, 2].

More and more companies, both big and small, are starting to understand the value and necessity of risk management [3]. According to some global studies, risk management significantly contributes to improving the productivity and value of organizations in aggressive changes of the society [4]. New directions in managing risks referred to that is required to continually examine the main business risks in the external environment.

Nowadays, using social media for business became essential as well as is gradually becoming a popular option for advertising businesses. It enables discussions to move beyond a private one-to-one discussion and into a discussion of many-to-many. Moreover, business owners may fully leverage features of social media for selling, promoting, and marketing in order to reduce costs. These features like sharing, tagging, chatting,

R. El Ayachi et al. (Eds.): CBI 2023, LNBIP 484, pp. 3–16, 2023.
https://doi.org/10.1007/978-3-031-37872-0_1

commenting, and notifying to advertise their goods, services, and brands. Since the cost is inexpensive and the amount of IT expertise required is modest, SMEs can use it for their daily transactions [5]. Further, social media can help the owners of firms to increase the productivity and competiveness. Also, risk management systems based on social media brings a new trend to identify and handle risks [6].

This study suggests an innovative system to identify and manage risks in SMEs based on social media perceptions using a lexicon approach. In best of our knowledge, this is the first contribution to gain insights from social media to predict risks from the users' interactions for SMEs. Also, we drive some basic explicit formulas for analyzing, identifying and assessing risks. Besides, we study Book Nook which is considered a SMEs in order to identify and assess potential risks in the tweets of their clients. Furthermore, some performance measures are calculated to validate the obtained results.

The rest of the article consists of: in Sect. 2 surveying the related articles related to this study. In Sect. 3, the suggested model is presented. In Sect. 4, the model is expressed in mathematical form and some important relations are driven. In Sect. 5, we study a practical use case. Future work and conclusion is included in Sect. 6.

2 Literature Review

Rozsa et al. 2021 [7], studied the perceptions of Corporate Social Responsibility impact from SMEs managers and owners on personal risk management in different countries by using online survey. The results demonstrated that people are major considered to be the basic company capital; thus, the entrepreneurs are more hopeful about the concentration of personal risk. In [8], they studied the present state of SMEs cyber risk management process to gain major insights for the future research by carrying out surveys. The results referred to lacks in risk culture and the exerted market for IT technical are issues, so investigation a relationship between the culture of cyber security and cyber risk management and would be available.

Asgary et al. 2020 [9] explored the method of perceiving main global risks in SMEs and described the effect of attributes and circumstance of country on SME risk evaluation and rankings as well as the difference from the global rankings by conducting online survey. The findings showed that the most effected risks on SMEs in ranking are universal economic risks and geopolitical risks, and the environmental risks are lowest in priority. The method of considering and managing in entrepreneurial SMEs, and the implications of integrating risk management in decision making approach are explored by utilizing a new model (RM-DM) [10]. The findings indicated that the proposed model improves handling risks and helps in making efficient decisions.

In [11], they aimed to considerably measure the block chain financial market risk and proposed appropriate prediction model for financial market in china by employing the meta-heuristic algorithm. Results proved that the efficiency of the new KMV model in identifying financial market risks such as credit risk in China. A conceptual framework on SMEs is proposed to assess commercial risk and provide management solutions by deploying machine learning [12]. Results concluded that the suggested framework helping SMEs from the perspective of ecosystem and liberate SMEs from elevated investments into self-growing human resource development.

Hudáková and Masar 2018 [13] recognized the common risks in Slovakia SMEs and their sources based on some factors such as the size of company and the period of doing business by using empirical research and statistical methods. The results showed that the significant of deploying risk management tools in SMEs that implies to progress in business performance. In [14], they examined the business risks in SMEs in Zilina region with more focusing on market risk and providing enterprise risk management approach for effective risk handling by conducting a survey. The results showed that most vital risks in the SMEs of Slovakia include market risk specifically in the case of insufficient managing for it.

Brustbauer 2016 [15] analyzed the risk management in SMEs by proposing a structural approach depending on questionnaire survey. The findings indicated that applying enterprise risk management can help SMEs to adapt with environment changes hence, enhancing competiveness and success in business. The main challenges in SMEs are examined according to risk management facility building and provided social capital role that probably help the risk management knowledge in founding and transferring facilitation by using cognitive capital-based RM model [16]. The findings concluded that the cognitive capital role is significant in SMEs for risk management facility building as well as cognitive capital is important in assembling structural and relational capital.

In [17], they focused on providing a comparative study to handle risk management in SMEs by employing a balance scorecard model. The results illustrated that the major improvement in SMEs that emerged this model more than the others. The present state of risk management in the SMEs of German is explored by providing a multidimensional scoring approach based deep analysis of questionnaire survey and provided a novel topology of risk management patterns [18]. The findings included that a recommendations could give to a firm in terms of which actions should the firm takes to enhance its risk management. Iskanius 2009 [19] discovered and evaluated the essential risk in Enterprise Resource Planning projects in terms of manufacturing SMEs by deploying company-specific risk analysis method. The findings gave recommendation of how to separate the ERB projects to easy sup projects.

Onwubolu and Dube 2007 [20] surveyed the most common used software tools in risk management with their main features and ranked them according to the features available. Also, they focused on presenting one of those tools through short case study. The results illustrated that deploying appropriate risk management software tools probably increase the performance and minimize costs in SMEs. The current patterns of risk management in the SMEs of German are examined and the linkage between risk management with the planning activities of business by using questionnaire survey [21]. The findings demonstrated that the controlling of risks depends on owners-managers and a few of SMEs deploying planning systems that the relationship between them and risk management is not well developed. The summary of the literature review is given in Table 1.

Table 1. Survey on risk management in SMEs.

Author/s	Year	Findings	Approach	Topic
Rozsa et al	2021	People are major considered to be the basic company capital; thus, the entrepreneurs are more hopeful about the concentration of personal risk	Online survey	Risk Identification
Hoppe et al	2021	Lacks in risk culture and the exerted market for IT technical are issues, so investigation a relationship between cyber risk management and cyber security culture would be available	Survey	
Asgary et al	2020	The most effected risks on SMEs in ranking are universal economic risks and geopolitical risks, and the environmental risks are in the bottom of ranking	Online survey	
Song et al	2020	The efficiency of the new KMV model in identifying credit risk in block chain financial market of china	Particle swarm optimization algorithm	
Hudakova et al	2017	Market risk is one of the most vital risks in the SMEs of Slovakia especially in the case of insufficient managing for it	Survey	

(continued)

Table 1. (*continued*)

Author/s	Year	Findings	Approach	Topic
Crovini et al	2020	The proposed model improves handling risks and helps in making efficient decisions	New model (RM-DM)	Risk Assessment and Control
Žigienė et al	2019	The proposed framework helping SMEs from the perspective of ecosystem and liberate SMEs from elevated investments into self-growing human resource development	Machine learning	
Hudáková, M., & Masar, M	2018	The significant of deploying risk management tools in SMEs that implies to progress in business performance	Statistical methods	
Brustbauer	2016	Applying enterprise risk management can help SMEs to adapt with environment changes hence, enhancing competiveness and success in business	Structural model depending on questionnaire survey	
Gao et al	2013	The cognitive capital role is significant in SMEs for risk management facility building as well as cognitive capital is important in assembling structural and relational capital	Cognitive capital-based RM model	

(*continued*)

Table 1. (*continued*)

Author/s	Year	Findings	Approach	Topic
Yiannaki	2012	The major improvement in SMEs that emerged this model more than the others	Balance scorecard model	
Henschel	2010	Recommendations could give to a firm in terms of which actions should the firm takes to enhance its risk management	Questionnaire survey	
Iskanius	2009	Recommendation of how to separate the ERB projects to easy sup projects	Company-specific risk analysis method	
Onwubolu et al	2007	Deploying appropriate risk management software tools probably increase the performance and minimize costs in SMEs	Short case study	
Henschel	2006	The controlling of risks depends on owners-managers and a few of SMEs deploying planning systems that the relationship between them and risk management is not well developed	Questionnaire survey	

3   The Framework of the Suggested Model

In this section, we give a novel framework for managing risks in SMEs based on social media indicators as in Figure 1. It describes the required stages for risk management process starting from risks identification then risks assessment then risks control and finish with the stage of risks monitoring.

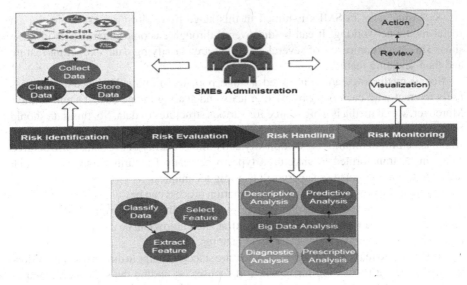

Fig. 1. Business intelligence risk management framework for SMEs.

According to the fast spread of using social media platforms by many SMEs around the world to conduct with their customers and search for new clients. To identify potential risks for SMEs, behavior of customers and opinions about their products should be studied. Other indicators for possible risks are the local, regional and global changes and the users discussions about them would be important in risks identification process for SMEs. This phase associated with three steps that are, collecting data, cleaning data and storing data. In this step, data can be collected form several social media platforms by using various data collection tools. Moreover, the data can be gathered from SMEs profiles on social media or from users profiles and their discussions and interactions on SMEs profile, local, regional and global changes. Many data collection tools such as APIs and crawlers and so on can be used to collect texts, images and videos and different structures of data. After collecting data, the gathered data from social media platforms will be cleaned from dirty data.

Cleaning process includes converting the dirty data to an acceptable data that leads to powerful insights for SMEs administration. The dirty data as deceptive news, missing data and incorrect data. Many of methods and techniques are suggested to control dirty data as in [22].

After cleaning data, the valid data will be stored in big data storage systems to be ready for further processing. These systems allow accessing and managing large-scale datasets effectively. For example, Cloud computing is considered a proficient storage facility for huge amount of data. It provides storage service in data centers with scalability, high availability, low latency and reduced costs. To illustrate, Amazon provides a storage service that allows customers to obtain unlimited storage on demand. Microsoft Azure, Google Cloud Storage and Dropbox are other examples for cloud storage facilities that based on the principle of storage-as-a service. Moreover, many several tools can be used for storing and manipulation big data such as Spark, Hive and NoSQL Databases.

Assessing risks in SMEs is aimed in this stage by exploring users' behavior on social media networking. It can be discovered through customer interactions in these sites. This stage comprises of several steps that are classifying data, extracting feature and then selecting feature.

Social media sites are considered a huge pool for multimedia big data whereas users interact with each other by writing texts and sharing images, videos and sounds. Moreover, social media is a big source for various structures of data. So, this data should be classified that depending on splitting the data to several homogeneous categories which is helpful in minimizing complexity and dimensionality of big data. Unstructured data can be transformed to structured type to be ready for further analysis through different techniques such as supervised learning techniques.

Based on the previous step, features extraction process can be done for each homogenous type of data and the output of this process will be collected in a group of all extracted features for the forthcoming process. The extracted features can be categorized into two types: linguistic features and social media features.

In addition to features extraction step, feature selection helps in dimensionality reduction and efficient processing for big data. It focuses on finding the significant features in the extracted feature set and ignore the irrelevant features that may lead to less accuracy in the analysis step. It can be done manually and by many methods such as Filters and Wrappers. Many other techniques can be used such as Embedded, Hybrid Method, Ensemble method and Integrative method.

The characteristic of analytics in big data can be divided into four types, namely, descriptive, diagnostic, predictive, and prescriptive. The descriptive models may assess, identify, and classify different connections and interactions in data. The diagnostic analytics provide answers to inquiries like "Why did it happen?" This analytics carefully examines data to identify the exact behaviour and origins of events. Data is transformed by predictive analytics into valuable and useful information. In order to predict future events, predictive analytics uses a variety of statistical techniques, including as modelling, machine learning, and game theory. Similar to predictive analytics, prescriptive analytics can suggest ways to make decisions about seizing an opportunity or lowering risks in the future. This analytics shows the results of each possible decision. In practice, prescriptive analytics may continuously and impulsively analyze fresh data to improve precision and provide wise decision-making options.

In monitoring phase, the previous processes and the obtained results are visualized by many tools to help the decision makers in understanding and capturing the insights easily.

4 Mathematical Model

Let $B_1, B_2, B_3, ..., B_n$ represent the customers or persons who are keen in a SMEs enterprise where n is the number of customers. They declare their conceptions about the enterprise on social media that give $P_1, P_2, P_3, ..., P_n$ of perceptions. The perception for any customer can be represented by the following vector of words

$$B_i = [w_1, w_2, w_3, ..., w_m], \tag{1}$$

where $1 \leq i \leq n$

The amount of gathered data can be huge and the words have several forms (nouns, verbs, adjectives, articles and prepositions). Hence, in order to decrease data size, data can be assorted multiple topics based on the following form:

$$T = \alpha N, \tag{2}$$

where N is the percentage of each noun in the whole perceptions and α is constant that equal one if the noun is relevant to the proposed application and zero in other case such that $T \geq 3\%$.

Therefore, the sample of buyers $B_1, B_2, B_3, ..., B_n$ can be categorized to similar groups based on Eq. (2). Also, another essential step in dimensionality reduction is to concentrating on the important words in sentence such as nouns, verbs, adjectives and adverbs and avoid the other forms such as conjugations, definite, indefinite, articles and so on. Thus, the vector in Eq. (1) convert to the following form:

$$B_i = [w_1, w_2, w_3, ..., w_\theta], \tag{3}$$

where $\theta < m$.

To be able to perform analysis for the customers' perceptions, we provide a rule that transform the major formats into numeric values based on the probability of referring to risk as shown in Eq. (4):

$$rule = \begin{cases} 0, & \text{no. crisk} \\ 0.5, & \text{potential .risk} \\ 1, & \text{risk} \end{cases} \tag{4}$$

Depending on the rule in Eq. (4), we establish an explicit formula to find the perceptions of each buyer as the following:

$$P_i = \beta \sum_{k=1}^{4} N_k, \tag{5}$$

where $\beta = (\frac{2-P}{2m})$ and m indicates to the number of words that refer to a risk in the context such that $1 \leq m \leq 4$.

So, the vector B_i will produce a vector which has the risk analysis values depending on Eq. (5). It has the following form:

$$g = [P_i]^T \tag{6}$$

For risk type recognition the Rt, we have:

$$Ri = R(sB), \qquad (7)$$

where R is has value one if there is word that refers to risk in the text and zero in other cases, B refers to the type mentioned risk in the sentence such as financial, market, operational risks and so on, and s is the associated value for each type of risk.

For risk measurement RM, we have:

$$RM = (R^2(1 + \sum_{k=1}^{2} C_k) - RP(1 + \sum_{k=1}^{2} C_k)), \qquad (9)$$

where P refers to the probability where $R \geq P$, C_1 and C_2 refers to maximum and minimum risks respectively, whereas C_1, C_2 and P equals one if any term in the text indicates to them and zero otherwise.

5 Experimental Results

Book Nook is one of top SMEs worldwide so, it is suitable as a use case for the suggested approach. More than 21 k tweets are collected depending on customers' geo-locations in the period from 01–02-2022 to 01–08-2022 by using Twint tool. Figure 2 shows the distribution of tweets in monthly:

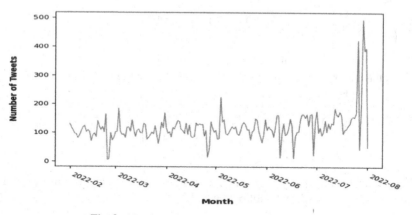

Fig. 2. Number of tweets distribution vs month.

By utilizing Eq. (2), the top frequencies of words in the dataset are shown in Fig. 3.

For analyzing the collected dataset based on finding risks or not, we will perform the risk analysis at the first. We focus on identifying risks in three common types of books namely, romance, fiction and story. The lexicon approach is used in analyzing process []. For romance books, we analyze more than 3.5 k tweets to recognize risks in each tweet. For fiction books, more than 1.6 k tweets as well as more than 890 tweets for story books are analyzed as in Fig. 4.

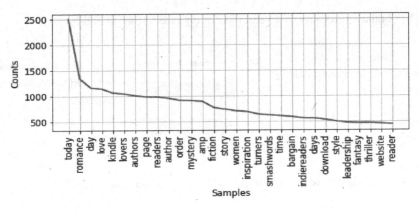

Fig. 3. Top frequencies of words in the sample.

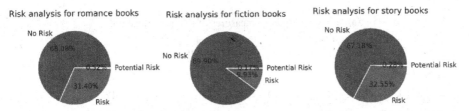

Fig. 4. Risk analysis for romance, fiction and story books.

Performance indicators as accuracy, precision, recall and f1-score are determined in Table 1. Three random samples from love, fiction and story books are selected and annotated (Table 2).

Table 2. Performance measures of risk analysis.

$N = 300$	Precision (%)	Recall (%)	F1-score (%)
No Risk	67	92	74
Risk	92	63	74
Potential Risk	100	70	82
Accuracy			75
Macro avg	85	75	78
Weighted avg	81	75	74

The obtained risks vary to several risks of types as shown in Fig. 5.

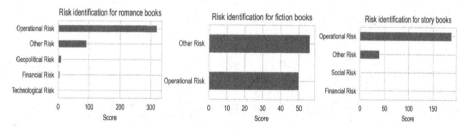

Fig. 5. Risk Identification for romance, fiction and story books.

Performance indicators for the risk identification is included in Table 3.

Table 3. Performance measures of risk identification.

N = 300	Precision (%)	Recall (%)	F1-score (%)
Operational risk	70	43	45
Financial risk	100	43	47
Geopolitical risk	100	50	47
Accuracy			46
Macro avg	72	50	48
Weighted avg	58	60	78

Besides, risk estimation in each category are provided in Fig. 6.

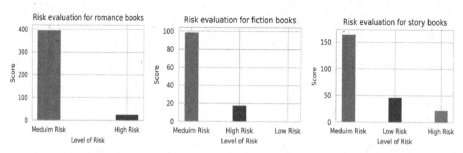

Fig. 6. Risk evaluation for romance, fiction and story books.

To validate the results of risk assessment process, Table 4 is given below:

Table 4. Performance measures of risk assessment.

$N = 300$	Precision (%)	Recall (%)	F1-score (%)
Low Risk	93	68	78
Medium Risk	100	68	81
High Risk	100	64	78
Accuracy			67
Macro avg	72	50	59
Weighted avg	95	67	78

6   Conclusion

This study proposed a novel business intelligence system to help managers of SMEs. This model depending on the perceptions of Twitter users and utilizing big data analysis techniques as well as natural language processing for data manipulation. In addition, we developed a mathematical formulas in explicit forms for analyzing, identifying and estimating risks. Besides, we gave a practical use case of Book Nook enterprise is studied for three categories namely, love, fiction and story books. Performance measures indicated to a good accuracy for the obtained results. In future work, other techniques such as machine learning and deep learning techniques will be employed to increase the efficiency and capability of the suggested model. To do that we will build dataset with labels to train these models and apply meta-heuristic algorithms to reduce the dimensionality of data.

References

1. Agarwal, R., Ansell, J.: Strategic change in enterprise risk management. Strateg. Chang. **25**(4), 427–439 (2016)
2. Klučka, J., Grünbichler, R.: Enterprise risk management–approaches determining its application and relation to business performance. Quality Innov. Prosperity **24**(2), 51–58 (2020)
3. Jankelová, N., Jankurová, A., Beňová, M., Skorková, Z.: Security of the business organizations as a result of the economic crisis. Entrepreneurship and Sustainability Issues **5**(3), 659–671 (2018)
4. Ribau, C.P., Moreira, A.C., Raposo, M.: SMEs innovation capabilities and export performance: an entrepreneurial orientation view. J. Bus. Econ. Manag. **18**(5), 920–934 (2017)
5. Derham, R., Cragg, P., Morrish, S.: Creating value: an SME and social media. PACIS Proceedings. Paper 53 (2011)

6. Darwiesh, A., El-Baz, A.H., Tarabia, A.M.K., Elhoseny, M.: Business intelligence for risk management: a review. American J. Bus. Operations Res. **6**, 16–27 (2022)
7. Rozsa, Z., Khan, K.A., Zvarikova, K.: Corporate social responsibility and essential factors of personnel risk management in SMEs. Polish J. Manage. Stud. **23**(2), 449 (2021)
8. Hoppe, F., Gatzert, N., Gruner, P.: Cyber risk management in SMEs: insights from industry surveys. The J. Risk Fin. (2021)
9. Asgary, A., Ozdemir, A.I., Özyürek, H.: Small and medium enterprises and global risks: evidence from manufacturing SMEs in Turkey. Int. J. Disaster Risk Sci. **11**(1), 59–73 (2020)
10. Crovini, C., Santoro, G., Ossola, G.: Rethinking risk management in entrepreneurial SMEs: towards the integration with the decision-making process. Manage. Decision (2020)
11. Song, Y., Zhang, F., Liu, C.: The risk of block chain financial market based on particle swarm optimization. J. Comput. Appl. Math. **370**, 112667 (2020)
12. Žigienė, G., Rybakovas, E., Alzbutas, R.: Artificial intelligence based commercial risk management framework for SMEs. Sustainability **11**(16), 4501 (2019)
13. Hudáková, M., Masar, M.: The assessment of key business risks for SMEs in Slovakia and their comparison with other EU countries. Entrepreneurial Bus. Econ. Rev. **6**(4), 145 (2018)
14. Hudakova, M., Schönfeld, J., Dvorský, J., Luskova, M.: The market risk analysis and methodology of its more effective management in SMEs in the Slovak Republic. Montenegrin J. Econ. **13**(2), 151–161 (2017)
15. Brustbauer, J.: Enterprise risk management in SMEs: towards a structural model. Int. Small Bus. J. **34**(1), 70–85 (2016)
16. Gao, S.S., Sung, M.C., Zhang, J.: Risk management capability building in SMEs: a social capital perspective. Int. Small Bus. J. **31**(6), 677–700 (2013)
17. Yiannaki, S.M.: A systemic risk management model for SMEs under financial crisis. Int. J. Organizational Analysis (2012)
18. Henschel, T.: Typology of risk management practices: an empirical investigation into German SMEs. Int. J. Entrep. Small Bus. **9**(3), 264–294 (2010)
19. Iskanius, P.: Risk management in ERP project in the context of SMEs. Eng. Lett. **17**(4) (2009)
20. Onwubolu, G.C., Dube, B.C.: Production planning & control: the management of operations. Prod. Plann. Control **17**(1), 67–76 (2007)
21. Henschel, T.: Risk management practices in German SMEs: an empirical investigation. Int. J. Entrep. Small Bus. **3**(5), 554–571 (2006)
22. Darwiesh, A., Alghamdi, M., El-Baz, A.H., Elhoseny, M.: Social media big data analysis: towards enhancing competitiveness of firms in a post-pandemic world. J. Healthcare Eng. (2022)
23. Elhoseny, M., Darwiesh, A., El-Baz, A.H., Rodrigues, J.J.: Enhancing cryptocurrency security using AI risk management model. IEEE Consumer Electronics Magazine (2023)
24. Darwiesh, A., El-Baz, A.H., Abualkishik, A.Z., Elhoseny, M.: Artificial intelligence model for risk management in healthcare institutions: towards sustainable development. Sustainability **15**(1), 420 (2023)

A Systematic Analysis for Machine Learning Based Cow Price Prediction

A. K. M. Tasnim Alam[1]([✉]), Zahid Hasan Nirob[1], Afrin Jahan Urme[1],
Rashidul Hasan Hridoy[1], Md. Tarek Habib[2], and Farruk Ahmed[2]

[1] Daffodil International University, Dhaka, Bangladesh
{tasnim15-12239,zahid15-12438,afrin15-12209,
rashidul15-8596}@diu.edu.bd
[2] Independent University, Bangladesh, Dhaka, Bangladesh

Abstract. Cow meets a significant number of demands for meats in South Asian countries, and a huge number of cows were sold in Bangladesh on the eve of Eid al-Adha. Cow prices depend on several factors, and determining the price of a cow is a cumbersome task for an inexperienced individual. Nowadays machine learning algorithms are significantly used for accurate price estimation. This study presents an efficient and accurate tool for determining cow prices using several characteristics of cows based on machine learning. Sixteen characteristics of a cow are considered in this study to determine its price. Four different machine learning algorithms were used and evaluated in this study for generating an accurate price prediction model. In this machine learning based systematic analysis, several experimental studies are conducted for evaluating models more precisely. Performance matrices were also used to evaluate machine learning algorithms.

Keywords: Systematic analysis · cow price prediction · machine learning · gradient boosting regression · decision tree regression · random forest regression · linear regression

1 Introduction

A remarkable number of people are engaged in cow farming, which also plays a crucial role in the economy of Bangladesh. Cow farming is also increasing day by day due to its demand. A cow is an essential source of food, nourishment, much-needed cash, and nitrogen-rich dung for use in regenerated soils and other applications. They also play a variety of roles in society. The main reason cows are raised as livestock is for their hides, which are used to make leather, milk, and food. Therefore, it follows that it is essential for everything to understand the value of a cow. Price prediction is difficult due to the intricacy of price fluctuation in the market. It also has a significant socioeconomic value [1].

However, in Bangladesh, where we are from, the majority of people live in rural areas. Since they live in villages, agriculture is their main source of income. Cows are the animal that is domesticated in the village's homes the most frequently as a result.

R. El Ayachi et al. (Eds.): CBI 2023, LNBIP 484, pp. 17–28, 2023.
https://doi.org/10.1007/978-3-031-37872-0_2

However, Bangladesh, a country where Muslims make up the majority, has a high demand for cows especially, in Eid al-Adha [2]. And also dairy cows have a big economic impact. Cows have a big influence on the economy. In the last ten years, the cost of beef has climbed by Bangladeshi Taka (BDT) 390 per kg, according to the TCB's market pricing list. The beef was priced per kilogram at BDT 275 in June 2013. Additionally, it was sold this year in June for 665 BDT. About 90 million and 83 thousand animals were sacrificed in the previous year of 2021. There were 4 million and 53 thousand cows among them. But in 2022 the number of animals taken to market for sacrifice is approximately 1 crore and 2.1 million, and 4.6 million 11 thousand of them were cows [1]. Figure 1 represents the increase in cow prices over the years, which strongly validates the increasing demand for cows.

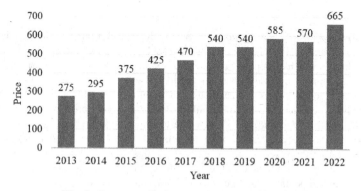

Fig. 1. Increase of price over the years (BDT/Kg)

The demand for cows is growing every year and is getting stronger every day [3]. Those who are unfamiliar with the price of cows are regularly taken advantage of while purchasing and selling them. Knowing the price of cows before selling or buying may be advantageous as a result, and the possibility of fraud will be reduced. Price prediction powered by machine learning is one of the most powerful tools [4]. Experts in machine learning have previously done a substantial amount of work on price forecasts, including stock and property price predictions, but cow price predictions are quite uncommon. As a result, the goal of this work is to predict how prices using a range of important factors. The price of cows will be predicted in this study utilizing a variety of attributes as well as algorithms. A certain number of people who are unaware of the price of cows will benefit from this and can buy cows at a predicted price. We used four machine learning algorithms such as random forest regression (RFR), decision tree regression (DTR), gradient boosting regression (GBR), and linear regression (LR). The key contributions of this study are summarized below:

- A new dataset is developed which contains several qualities of cows with real prices.
- An efficient machine learning-based framework is developed for accurate cow price prediction, which is evaluated using several experiments.

The remainder of this essay is structured as follows: It is covered in Section 2. Section 3 describes the details of the research methodology followed in this study. The outcomes obtained in this paper are provided and discussed in Section 4. Finally, this paper is concluded in Section 5.

2 Literature Review

Several studies were conducted in the literature using machine learning for predicting the prices of several goods. Rahman et al. researched predicting cow pricing ranges by using any cow photograph [5]. Cow photographs were gathered from a variety of online e-commerce sites that sell cows to primarily anticipate the price range of cows based on the photos of the cows. A convolutional neural network (CNN) is utilized as an image classifier and linear regression is applied to estimate pricing in a machine learning-driven method of prediction. An F1 score of 65.73%, 70.05% accuracy, 68.69% precision, and 70.05% recall are assigned to this piece of work. The findings indicate that a cow's price range may be forecasted with 70% accuracy. Lawrence et al. analyzed the cattle price forecast errors based on live cattle futures and seasonal index [6]. Overall, the seasonal index's predictions performed somewhat better than the Futures, with an average inaccuracy of -.1% as opposed to.7%. The average forecast inaccuracy among researchers predicts that the actual price will typically be.4% higher than the basis-adjusted Futures price. Bozic et al. works on testing for futures pricing and implied volatility biases using parametric bootstraps to rate livestock margin coverage for dairy cattle [7]. For testing the presence of bias in futures prices and implied volatility in deferred contracts with overlapping time-to-maturity horizons, the authors of this paper have developed a parametric bootstrap method. They use their methodology to test the hypotheses that futures prices are accurate and impartial predictors of terminal prices and that implied volatility squared times remaining time to expiration is an accurate predictor of terminal log-price variance. They employ the LGM-Dairy rating method, which is predicated on the unbiasedness of implied volatilities derived from at-the-money options and futures prices.

Eldridge et al. have used included two distinct yet closely related components of the decision tool for grazing management and profitability predictions and the pricing analysis of feeder cattle [8]. Kinnischtzke et al. analyzed identifying key, easily accessible features of the beef supply that influence changes in feeder and slaughter cattle prices [9]. Data were derived from the USDA's quarterly and monthly cattle-On-Feed data. Muwanga et al. [10] conducted relying on a few U.S. cattle marketplaces, this study examined the spatial linkages, price distributions, and price forecasts for cattle. In this research, the characteristics of distinct cattle price series, the application of a rational expectations model to cattle pricing for various marketing locations and cattle classes, and the spatial price correlations for certain markets were all investigated. Marsh et al. utilized a rational distributing lag method to forecast quarterly live cattle prices [11]. Quarterly U.S. feeder cattle and fed prices for cattle were determined inside a theoretically distributed lag framework to minimize problems with specification errors in the disturbance structure. These dynamics were found to perform better than static-serial correlation and fully autoregressive requirements.

Franzmann et al. developed Feeder, the author, and Wholesale Beef Cattle Price Trend Models [12]. Harmonic regressions were applied to monthly data to provide a low-cost alternative technique of forecasting prices at three market levels in the cattle industry and to provide insight into the historical connections among these market levels. Coffey et al. examined the impacts of alterations in price momentum and market factors on hedging live cattle [13]. The effect of shifting fundamental economic variables and price trends on the predictability of the basis for live cattle is quantified in this study. To understand how changes in the live cattle market and current market conditions affect basis predictability, basis prediction errors are modeled as functions of pertinent economic variables. Moser et al. measured the precision of direct genetic values in Holstein bulls and cows was assessed using certain subsets of SNP markers [14]. This study's goal was to assess how well low-density assays predicted direct genomic value (DGV) for two profit index characteristics, a survival index, an overall conformation trait, and five milk production traits (APR, ASI). For cows, subsets with 3,000 SNP offered more than 90% of the accuracy possible, and for young bulls, 80% of the accuracy was possible with a high-density test. Blakely et al. assessed a quarterly cat farm price projection model intending to develop a futures market approach [15]. According to the structural model, the inventory of heifers, steers, and bulls weighing less than 500 lb, the price of maize, the price of fed cattle, and the price of fed cattle with a two-period lag are some of the more significant variables in determining the price of feeder cattle. Analytically reduced forms have low usefulness in predicting the price of feeder cattle; instead, it would appear preferable to create an estimate of the price of feeder cattle using a straightforward OLS model. Although the futures market would seem to be the ideal source for such a series, the database is currently too small to include futures prices in a function that forecasts the price of feeder cattle. A few straightforward tactics that combine futures markets with price projections show promise for both lowering the revenue variance that would be observed under a cash-only business and increasing the average income obtained by feeder cattle producers. It appears encouraging to conduct more research in the field of developing strategies.

3 Methodology

This study's primary goal is to create a model for estimating the price of a cow using machine learning. The price of a cow is affected by several aspects. To prepare our dataset, we must use the preprocessing method and carry out data exploration. The creation of a dataset from scratch and the use of models were the challenging aspects of the study procedure. It could be difficult to select the model that best fits the dataset. Figure 2 illustrates the phases of our methodology.

The methodology part is mostly divided into two sections. In the first, data analysis is covered, and in the second, machine learning model development and evaluation are covered. Each of these provides a detailed explanation of the method for making predictions, including the methods used to collect, prepare and study data for data analysis. Additionally, the remaining portion includes information on the training procedure, suitable predictive models, a prediction of cow prices, and accuracy testing for machine learning evaluation.

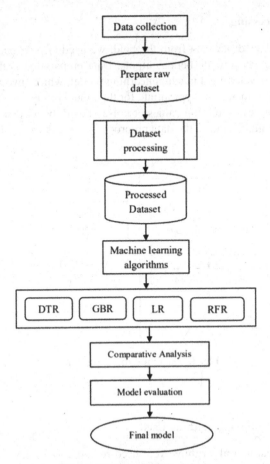

Fig. 2. Phases of Methodology

3.1 Data Collection

Data is an essential part of any machine learning system. We intend to focus on the cows that are offered on the regular market when implementing the system. Due to the regular fluctuation in the price of cows, the collecting of updated data is necessary. Data was gathered from the cow dealers, regular market vendors, and customers. The main emphasis of the data collection process was the Eid al-Adha event. Our primary data collection criteria included weight, height, gender, color, horn status, age, teeth, bread, origin, and price. Three types of gender were taken into consideration Male, Heifer means a female that has not had any offspring, and female cow that had offspring before. In our data collection, there were 11 different breeds of cows. Based on sixteen features, a thousand pieces of data were gathered.

3.2 Data Preprocessing

Since we have collected our data from the field, we need to preprocess the dataset so that the algorithms can provide better outputs. Data preparation is the first and most important stage in developing a machine learning model, which involves preparing the unprocessed data and making it appropriate for the model because real-world data is in an improper format or has missing values or noise, it can't be used straight in machine learning models. Figure 3 shows the data preprocessing steps we used.

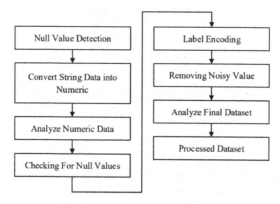

Fig. 3. Data Pre-processing steps

3.3 Machine Learning Algorithms Used

Several machine learning algorithms were used remarkably to develop efficient price prediction models for several goods. LR, STR, RFR, and GBR performed significantly better than other algorithms in works introduced in the literature.

LR: One of the most fundamental and well-known Machine Learning algorithms is LR. It is a method of analysis that is employed in predictive analysis. Predictions are made using linear regression for continuous/real/numeric variables like humidity, temperature, rainfall, and productivity, among others. There are two varieties of linear regression, and multiple regressions utilized in this study [16]. Here, the Dependent variable is identified by y. X represents an independent variable. m stands for "Estimated Slope." b refers to the estimated intercept. The LR equation is shown in Eq. 1 below.

$$y = mX + b \qquad (1)$$

GBR: Applications of a machine learning method called gradient boosting include classification and regression tasks. It provides a prediction model by combining inaccurate estimation techniques like decision trees. When a DT is a poor learner, the ensuing method, known as gradient-boosted trees, typically outperforms random forest. A gradient-boosted trees model is constructed using the same stage-wise design method as prior boosting strategies, however, this method improves on earlier methods by allowing the optimization of any discrete weight vector [17]. The GBR equation is shown in Eq. 2 below.

$$F(x) = \sum_{m=1}^{M} \gamma_m h_m(x)$$
(2)

Here, $F(x)$ is the mean anticipated target value for input x in this case. The average number is M. The learning rate or mean weight is m. The weak learner's average output for input x is represented by $h_m(x)$.

DT: A DT is a technique for decision analysis that uses a paradigm similar to a tree to describe options and their results. In contrast to other supervised learning approaches, the decision tree technique can handle both classification and regression tasks. By analyzing fundamental decision rules produced from a set of data (training data), a decision tree is used to construct a training structure that may be used to forecast the class or value of the target variable [18]. Here, the input x's mean anticipated target value is represented by $F(x)$. The average number of leaf nodes is M. The m th leaf node's associated mean constant value is cm. Indicator function $= I(x \epsilon R_m)$. The DT equation is shown in Eq. 3 below.

$$F(x) = \sum_{m=1}^{M} c_m . I(x \epsilon R_m)$$
(3)

RF: DTs are constructed in a relatively large quantity during the training phase of the random forest technique, which is used for classifying, regression, and other applications. The class that the majority of the branches in a classification challenge chose is the outcome of the random forest. The average prediction made by a particular tree is provided for regression tasks. Random forests are a better fit for this since decision trees frequently overfit their trained model. Random forests usually outperform decision trees, but they are less precise than gradient-enhanced trees. However, their effectiveness of them may be impacted by data properties [19]. Here, $F(x)$ is the mean anticipated target value for input x in this case. There are N total decision trees. $f_i(x)$ is the decision tree's estimated target value for the input x. The RF equation is shown in Eq. 4 below.

$$F(x) = \frac{1}{N} \sum_{i=1}^{N} f_i(x)$$
(4)

3.4 Performance Metrics Used

The performance of the algorithms was assessed using the three most popular accuracy measures for regression models: Mean Absolute Error (MAE), Mean Squared Error (MSE), and Root Mean Square Error (RMSE).

MAE: In the sense of machine learning, absolute error is the magnitude of the difference between the expected value of an observation and its actual value. Utilizing the mean of absolute errors for a group of estimates and observations, MAE calculates the size of errors for the entire group.

The absolute error is the result of calculating the prediction error by deducting the actual value from the predicted value. MAE, which is the average of all absolute errors. Statistically, MAE refers to the results of measuring the difference between two continuous variables [20].

Here, The number n indicates the total amount of samples or observations. The letter i is used to indicate each distinct forecast. The letter y indicates the equivalent true or observed value. The symbol for the summation operator denotes the addition of all the absolute variations between the expected and actual values. The MAE equation is shown in Eq. 5 below.

$$MAE = (1/n) * \sum |i - y| \tag{5}$$

MSE: The mean squared error tells us how much a regression line reflects the collection of data points. It is an uncertain function that relates to the anticipated value of the squared error loss. MSE is calculated as the average, more specifically the mean, of errors squared from values about a function. We square the variation between our model's predictions and the actual data, average it across the entire dataset, and then calculate the MSE [21].

Here, N is the total sample quantity of observations, while i denotes each unique prediction. y_i represents a comparable true or observed value. The symbol for the summation operator denotes the addition of any squared discrepancies between the actual and anticipated values. The MSE equation is shown in Eq. 6 below.

$$MSE = \frac{1}{N} \sum_{i=1}^{N} \left(y_i - \hat{y}_i \right)^2 \tag{6}$$

RMSE: One of the methods for assessing the accuracy of forecasts is the RMSE. It displays the Euclidean distance between predicted values and measured actual values. For every point of data, determine the residual (difference between predictions and truth), its norm, the mean, and its square root to determine the residual mean square error (RMSE). Since RMSE requires and uses correct measurements at each projected data point, supervised learning algorithms frequently use RMSE. The RMSE is formally defined as the equation shown in Eq. 7 below.

$$RMSE = \sqrt{\frac{\sum_{i=1}^{N} \|y(i) - \hat{y}(i)\|^2}{N}} \tag{7}$$

where N refers to several data points, $y(i)$ means the ith measurement and $\hat{y}(i)$ is the prediction that corresponds to it [22].

By quantifying the discrepancy between actuals and estimates, the RMSE highlights the existence of outliers. We used RMSE as a critical, essential measure to assess the efficacy of machine learning models. The MSE calculates a prediction approach's degree of accuracy. It evaluates the average square root of the discrepancy between both the actual and predicted values. The MSE is equal to zero whenever a model is error-free. It can be contrasted with models whose mistakes are computed in different units. Although MAE calculates the absolute mean difference between the actual and expected data, it does not permit appreciable error rates. The square means the difference between both the estimated and real data is calculated via MSE.

4   Experimental Evaluation

This study covers the price prediction model of cows. As per our best knowledge, no study in conducted in the literature for determining the prices of cows using machine learning. There were 1000 total pieces of data collected. The primary sources of data were cow markets or cow traders. We used four machine-learning algorithms to find out the best approaches to predict better price accuracy. Finding the most effective method for this task required an in-depth investigation of many methods which was a critical component of this study. We used GBR, DTR, RFR, and LR to develop a price prediction model. Performance matrices were used to evaluate machine learning as well. In our dataset, we used real cow prices, and all values obtained by MAE, MSE, and RMSE were divided by 1000. We discussed MAE, MSE, and RMSE values with the help of the currency of Bangladesh, which is BDT.

The lowest MAE, roughly 6.8 BDT, was attained via LR. Though it may appear to be a significant sum overall, when compared to cow prices, it is a little sum. The highest number, 11.5 BDT, was reached by DTR in contrast. The MAE of GBR is a little higher than that of LR. On the other hand, The MAE of the GBR is lower than the RFR's by quite a margin. The experimental performance for MAE is shown below in Fig. 4.

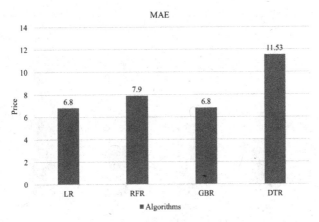

Fig. 4. The experimental performance of MAE

When compared to other algorithms, LR produced the lowest MSE result, which is 223943 BDT. And DTR generated the highest output. In addition, the MSE of the GBR is smaller than that of the RFR. The MSE of GBR is lower and closest to that of LR. The experimental performance for MSE is shown below in Fig. 5.

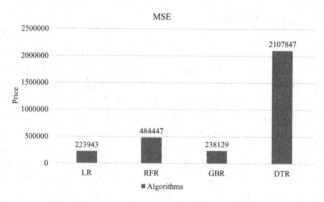

Fig. 5. The experimental performance of MSE

Similar to the earlier error calculations, LR had the lowest RMSE, which is around 14.96 BDT. The closest to LR is GBR, around 15.4 BDT. The highest number was obtained by DTR, which is 45.9 BDT. The experimental performance for RMSE is shown below in Fig. 6.

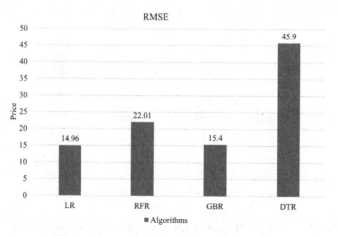

Fig. 6. The experimental performance of RMSE

5 Conclusion and Future Work

Our primary objective in this paper was to use machine learning to forecast the price of a cow. On this foundation, we gathered our data taking into consideration all the characteristics necessary to forecast the price of cows. To accurately anticipate the price of cows, we have undertaken extensive studies utilizing a variety of machine-learning algorithms. During the course of this research, four machine-learning algorithm principles were applied. Which are LR, DTR, RFR, and GBR. With the use of performance matrices like MAE, MSE, and RMSE, we also evaluate these techniques. LR achieved the lowest score in the calculations for the three MAE, MSE, and RMSE errors. And when compared to other algorithms, GBR's score was the closest to LR.

In our modern technology machine learning makes our life very faster, easier, and more comfortable in every sector of human life. In the future, we will try to extend our dataset. Machine learning works extremely very well with large amounts of data, a large amount of dataset should be collected. Other machine-learning techniques can be utilized for creating a model that is more efficient and intelligent. We hope our research will help people to know the price of a cow and it will help people to purchase a cow easily.

References

1. Predicting feeder cattle prices. https://www.alberta.ca/economics-and-marketing-predicting-feeder-cattle-prices.aspx
2. Cattle traders expect good business this Eid ul Azha. https://www.thedailystar.net/business/economy/news/cattle-traders-readying-eid-market-3048601
3. Livestock Industry of Bangladesh: Growth, Challenges and Future Opportunities. https://businessinspection.com.bd/livestock-industry-of-bangladesh/
4. Price Prediction: How Machine Learning Can Help You Grow Your Sales. https://dlabs.ai/blog/price-prediction-how-machine-learning-can-help-you-grow-your-sales/
5. Rahman, M.A., Kabir, M.A., Haque, M.E., Hossain, B.M.M.: A machine learning-based price prediction for cows. In: 2021 American Journal of Agricultural Science, Engineering, and Technology 5(1), 64–69 (2021)
6. Lawrence, J., Hoffman, C.: Cattle Price Forecast Errors: Live Cattle Futures and Seasonal Index. Futures, pp. 2000–2008 (1990)
7. Newton, J., Thraen, C.S., Bozic, M.: Evaluating policy design choices for the margin protection program for dairy producers: an expected indemnity approach. In: 2016 Applied Economic Perspectives and Policy, Agricultural, and Applied Economics Association, 38(4), pp. 712–730 (2016)
8. Eldridge, R.W.: Kentucky Feeder Cattle Price Analysis: Models for Price Predictions and Grazing Management, p. 173 (2015)
9. Kinnischtzke, J.N.: Forecasting cattle price trends. In: 1988 North Dakota State University ProQuest Dissertations Publishing, p. 1334941 (1988)
10. Muwanga, G.S.: Cattle price dispersions, spatial relationships, and price predictions: a case study of selected United States cattle markets. In: 1998 PhD diss, Utah State University (1998)
11. Marsh, J.M.: A rational distributed lag model of quarterly live cattle prices. In: 1983 American Journal of Agricultural Economics 65(3), 539–47 (1983)
12. Franzmann, J.R., Walker, R.L.: Trend models of feeder, slaughter, and wholesale beef cattle prices. In: 1972 American J. Agricultural Econ. 54(3), 507–12 (1972)
13. Coffey, B., Tonsor, G., Schroeder, T.: Impacts of changes in market fundamentals and price momentum on hedging live cattle. In: 2018: J. Agricultural and Res. Econ. 43, 18–33 (2018)
14. Moser, G., Khatkar, M.S., Hayes, B.J., et al.: Accuracy of direct genomic values in Holstein bulls and cows using subsets of SNP markers. In: 2010 Genet Selection Evol 42, 37 (2010)
15. Blakely, P.K.: A quarterly feeder cattle price forecasting model with application towards the development of a futures market strategy. In: 1976 PhD diss., Virginia Polytechnic Institute and State University (1976)

16. Varma, A., Sarma, A., Doshi, S., Nair, R.: House price prediction using machine learning and neural networks. In: 2018 Second International Conference on Inventive Communication and Computational Technologies (ICICCT), pp. 1936–1939, Coimbatore, India (2018)
17. Gumus, M., Kiran, M.S.: Crude oil price forecasting using XGBoost. In: 2017 International Conference on Computer Science and Engineering (UBMK), pp. 1100–1103, Antalya, Turkey (2017)
18. Machine learning classifiers – the algorithms and how they work. https://monkeylearn.com/blog/what-is-a-classifier/
19. Understanding Random Forest. https://www.analyticsvidhya.com/blog/2021/06/understandingrandom-forest/
20. Mean Absolute Error. https://c3.ai/glossary/data-science/mean-absolute-error/
21. Mean Squared Error: Overview, Examples, Concepts and More. https://www.simplilearn.com/tutorials/statistics-tutorial/mean-squared-error
22. Root Mean Square Error (RMSE). https://c3.ai/glossary/data-science/root-mean-square-error-rmse/

Deep Learning Models for Cybersecurity in IoT Networks

Meriem Naji[1]([✉])[ID] and Hicham Zougagh[2]

[1] Computer Sciences Department, Sultan Moulay Slimane University,
Beni-Mellal, Morocco
meriemnaji39@gmail.com
[2] Computer Sciences Department, Sultan Moulay Slimane University,
Beni-Mellal, Morocco
h.zougagh@usms.ac.ma

Abstract. The idea of the Internet of Things (IoT) was developed to im- prove people's lives by providing diversity range of intercon- nected smart devices and applications in several areas.

However, ensuring security in IoT environments remains a critical challenge, primarily due to the various security threats that devices face. While several approaches have been proposed to secure IoT devices, there is always room for improvement. One promising avenue is leveraging machine learning, which has shown its ability to identify patterns even in situations where traditional methods fail. Deep learning, in particular, offers an advanced approach to enhancing IoT security.

One transparent option for anomaly-based detection is the utilization of deep learning techniques. This article introduces several approaches based on Recurrent Neural Networks (RNNs) using Long Short-Term Memory (LSTM), Autoencoders, and Multilayer Perceptrons. By harnessing the power of IoT, these anomaly-based Intrusion Detection Systems (IDS) offer the capability to effectively analyze all traffic flowing through the IoT network. The proposed model exhibits the ability to detect any potential intrusions or abnormal traffic behavior. To validate its effectiveness, the model is trained and tested using the NSL-KDD datasets, achieving an impressive accuracy of 97.85% for binary classification and 97.98% for multiclass classification.

Keywords: IA · Network Intrusion Detection System · Deep Learning · Security · NSL-KDD

1 Introduction

The Internet of Things (IoT) has garnered significant attention in recent times, owing to its diverse applications and contributions in various fields such as industrial operations, healthcare, and smart environments [1]. IoT refers to a network of physical objects equipped with sensors, software, and connectivity, enabling them to communicate with other connected devices over the Internet. This capability to remotely monitor and control devices has given rise to numerous novel

R. El Ayachi et al. (Eds.): CBI 2023, LNBIP 484, pp. 29–43, 2023.
https://doi.org/10.1007/978-3-031-37872-0_3

applications across industries, including connected industrial systems, smart homes, health monitoring systems, and energy management systems [1,2].

The IoT has become an integral part of our daily lives, offering numerous objects that enhance convenience and practicality, such as virtual companions [3]. With the ever-growing telecommunications era, IoT adoption is rapidly increasing, connecting a wide range of devices. Recently, there has been a renewed interest in exploring the potential of deep learning technology to assess and enhance the security of digital devices [4,5]. Consequently, a significant amount of data is being generated to combat potential risks. Employing artificial intelligence (AI) in IoT is one approach to alleviate the heavy task load associated with various processes.

Despite the manifold services and applications provided by the IoT, it remains susceptible to cyberattacks. The primary concerns for IoT systems revolve around the security of physical devices and the protection of data against external threats and attacks. Cyberattacks involve the deliberate exploitation or unauthorized access to an individual's or organization's information or infrastructure. The diverse range of IoT devices and protocols, coupled with devices openly interfacing with the Internet and limited computing power, pose challenges in safeguarding IoT devices from attacks [6,7].

Due to the heterogeneous nature of the IoT ecosystem and the lack of interoperability, traditional security solutions prove ineffective [8]. However, other aspects of IoT security, such as maintaining data integrity, confidentiality, and secure user access, have seen improvements. Nonetheless, these security methods, although designed with users and the IoT ecosystem in mind, still have certain limitations.

Hence, a dedicated module is necessary to ensure IoT network security. One example of a security measure already employed in wireless networks is the intrusion detection system (IDS) [3,4]. Incorporating IDS features from wireless networks can aid in safeguarding the IoT network against attacks and other issues.

In this article, we will introduce a novel intrusion detection system based on deep learning techniques.

The Purpose of Work. The objective of this study is to implement several intrusion detection methods based on Deep Learning approaches and to evaluate the performance of the systems developed. We use the NSL-KDD dataset, which represents real network traffic containing several types of attacks (DOS, Probes, R2L, U2R).

2 Related Works

The primary goal of anomaly detection is to identify patterns in data that deviate from expected behavior. In our work, we can utilize anomaly detection techniques to differentiate between normal and attack traffic. Simple threshold-based methods often misclassify normal traffic as abnormal and struggle to adapt to evolving attack patterns [9]. To mitigate false positives, more advanced anomaly detection algorithms, particularly those based on machine learning, can be employed.

Deep neural networks, in particular, show promise in outperforming traditional machine learning techniques when applied to large datasets.

Anomaly detection has long been employed in network intrusion detection systems (NIDS) for detecting malicious activities in non-IoT networks. Therefore, insights from NIDS literature can guide the selection of anomaly detection methods for IoT networks. Nearest neighbor classifiers [10], support vector machines [11], and rule-based approaches like decision trees ánd random forests [12,13] have been identified as effective approaches in this context.

Although there are similarities between NIDS and IoT botnet detection, limited research has been conducted on adapting anomaly detection specifically to IoT networks.

In Deshmukh's work in 2014 [14], an IDS was developed using the Naive Bayes classifier with various preprocessing methods. The study utilized the NSL-KDD and WEKA datasets for experimental analysis. The results were compared with other classification algorithms such as NB TREE and AD Tree, indicating that Naive Bayes exhibited a lower execution time while maintaining a reasonable true positive rate.

In Noureldien Yousif's study in 2016 [15], the performance of seven supervised machine learning algorithms was evaluated for detecting Denial of Service (DoS) attacks using the NSL-KDD dataset. The experiments utilized the Train+20% file for training and the Test-21 file for testing, employing a 10-fold cross-validation to assess the methods' performance on unseen data.

Jabbar and Samreen presented a novel approach in 2016 [16], employing Alternating Decision Trees (ADTs) for attack classification, which is typically used for binary classification problems. Their proposed model demonstrated a higher detection rate and reduced false alarm rate compared to traditional IDS attack classification methods.

3   Proposed Methodologies

3.1   Dataset Description

The NSL-KDD intrusion detection dataset is derived from the well-known KDD Cup 99 dataset, which was created in 1999 for a machine learning competition. The objective of the competition was to accurately classify network connections into five categories: Normal, Denial of Service (DoS), Network Probe (Probe), Remote to Local (R2L), and User to Root (U2R). Each connection in the dataset consists of 41 attributes that enable classifiers to predict its appropriate class. These attributes encompass information and statistics derived from monitoring a simulated US local network in 1998, such as the connection duration, protocol type, percentage of connections for the same service, and more [17].

The NSL-KDD dataset was created in 2009 as an improved version of the KDD Cup 99 dataset, aiming to address some of its inherent issues [17]. It comprises the KDDTrain+ dataset as the training set and the KDDTest+ and KDDTest-21 datasets as the test sets, with each set containing different normal records and four different types of test data. Notably, the KDDTest-21 dataset is a subset of the KDDTest+ dataset and poses a greater challenge in terms of classification.

To enhance the dataset, the NSL-KDD dataset incorporates modifications to the original KDD Cup 99 data. Redundant or duplicated connections, which constituted a significant portion (75% to 78%) of the dataset, have been eliminated. Consequently, the total number of data instances has been significantly reduced, with the NSL-KDD dataset containing 125,973 connections compared to the 805,050 connections in the KDD Cup 99 dataset.

3.2 Data Preprocessing

Grouping of Attacks. In this phase, we categorized the attacks within the NSL-KDD dataset into four distinct groups, as illustrated in the following Table 1:

Table 1. Attacks categories in NSLKDD dataset

Attack Category	Description	Attack type
DoS	Denial of service (DoS) is a deliberate and malicious attack that aims to render a service unavailable, thereby preventing legitimate users from accessing it.	neptune, smurf, processtable, teardrop, udpstorm, back, land, pod, mailbomb, apache2, worm
Probe	A probe or surveillance attack is carried out with the intention of gathering crucial information about a network's security. The attacker aims to assess and exploit vulnerabilities in order to manipulate the security settings of the network.	ipsweep, nmap, portsweep, satan, mscan, saint
R2L	This class of attacks involves attempting to gain unauthorized local access to a remote machine by sending packets to the network. The attacker aims to exploit vulnerabilities or weaknesses in the network to gain privileged access to the targeted machine	ftp write, guess passwd, sendmail, named, snmpgetattack, imap, multihop, phf, spy, warezclient, warezmaster, snmpguess, xlock, xsnoop, httptunnel
U2R	The primary objective of this attack is to illicitly explore or steal data, install viruses, or cause harm to the victim by gaining unauthorized access using a normal user account. The attacker aims to exploit the compromised user account to carry out malicious activities without being detected or raising suspicion.	buffer overflow, rootkit, ps, sqlattack, Loadmodule, perl, xterm

Label Conversion and Feature Numericalization. In our classification process, we assigned a numerical value to each type of attack in the NSL-KDD dataset. Specifically, we used the following numerical assignments: 0 for Normal, 1 for Denial of Service (DoS), 2 for Probe, 3 for Remote to Local (R2L), and 4 for User to Root (U2R).

As most machine learning algorithms do not handle categorical data directly, it was necessary to convert the non-numeric features, namely 'protocol-type', 'service', and 'flag' in the NSL-KDD dataset. To accomplish this, we utilized the pandas library and applied the 'factorize()' method, which mapped the corresponding non-numeric features 'proto' and 'service' to numerical representations. This conversion enabled us to incorporate these features into the machine learning algorithms for further analysis and modeling.

Standardization. Standardization refers to a scaling technique that involves centering the data around its mean and adjusting it to have a unit standard deviation. It is a crucial step in data preprocessing, particularly when dealing with features that have widely varying ranges, such as time, source bytes, and destination bytes.

To achieve standardization, the StandardScaler algorithm, as described in [18], is commonly employed. The formula for standardization is given by:

$$z = \frac{x_i - \text{mean}(x)}{\text{stdev}(x)} \tag{1}$$

Here, $\text{mean}(x)$ represents the mean value of the training sample, $\text{stdev}(x)$ represents the standard deviation of the training sample, and x_i corresponds to the feature value being standardized. By applying this formula, each feature is transformed to have a mean of zero and a standard deviation of one, enabling fairer comparisons and facilitating the performance of certain machine learning algorithms.

Performance Indicators. In order to effectively evaluate the performance of the proposed intrusion detection method, several performance measures are adopted, including accuracy, precision, recall, and F1 score. These metrics are calculated based on the Network Attack Classification confusion matrix, as depicted in Fig. 1.

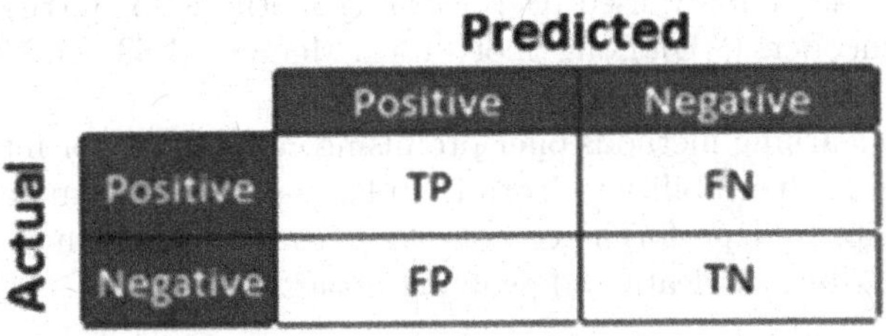

Predicted

		Positive	Negative
Actual	Positive	TP	FN
	Negative	FP	TN

Fig. 1. Confusion matrix for Network Attack Classification.

Accuracy (AC) represents the percentage of records that are correctly classified out of the total number of records, as shown in Equation (2).

$$Accuracy = AC = \frac{TP + TN}{TP + TN + FP + FN} \tag{2}$$

True Positive Rate (TPR), also known as Detection Rate (DR), indicates the percentage of anomaly records that are correctly identified out of the total number of anomalies, as expressed in Eq. (3).

$$TPR = \frac{TP}{TP + FN} \tag{3}$$

Precision is the number of correctly identified positive instances (TP) divided by the total number of instances classified as positive (TP + FP), as given by Eq. (4).

$$Precision = \frac{TP}{TP + FP} \tag{4}$$

Recall is the same as TPR and represents the number of correctly identified positive instances (TP) divided by the total number of actual positive instances (TP + FN), as shown in Eq. (4).

F1 score is the harmonic mean of precision and recall, providing a balanced measure between the two. It is calculated using Equation (5).

$$F1 = \frac{2 \times Precision \times Recall}{Precision + Recall} \tag{5}$$

These performance measures are essential for assessing the effectiveness and reliability of the intrusion detection method in accurately classifying network traffic and identifying anomalies.

3.3   Methodology

In recent years, there has been a growing interest in applying various intrusion detection methods, including machine learning, ensemble learning, and deep learning techniques.

Deep learning methods have gained popularity due to their ability to automatically extract relevant features and perform classification tasks in intrusion detection. Some commonly used deep learning methods for intrusion detection include: Autoencoders (AE), Long Short-Term Memory (LSTM), Multilayer Perceptron (MLP).

These deep learning methods offer promising capabilities for intrusion detection by leveraging their ability to learn from large-scale and complex data. They have shown improved performance compared to traditional machine learning methods in handling intricate and evolving attack patterns.

3.4 Long Short-Term Memory (LSTM)

Long Short-Term Memory (LSTM) is a type of recurrent neural network (RNN) architecture that is specifically designed to capture and model long-term dependencies in sequential data. It addresses the limitations of traditional RNNs by introducing memory cells and gating mechanisms that allow for better information retention and gradient flow.

Another variant of the RNN architecture is the Gated Recurrent Unit (GRU), which also aims to overcome the issues of vanishing gradients and information loss in long sequences. GRU simplifies the LSTM architecture by combining the forget and input gates into a single update gate and merging the cell state and hidden state.

Both LSTM and GRU architectures are capable of modeling sequential data and capturing long-range dependencies. However, there are some differences between them. GRU typically has a simpler structure and requires fewer parameters compared to LSTM, making it more memory-efficient and faster to train. On the other hand, LSTM is known to perform better in datasets that involve longer sequences, as it has more capacity to store and retain information over extended periods.

The choice between LSTM and GRU depends on the specific requirements of the task at hand, such as the length of the sequences and the available computational resources. Researchers and practitioners often experiment with both architectures to determine which one yields the best results for a given dataset or problem domain [19] (Fig. 2).

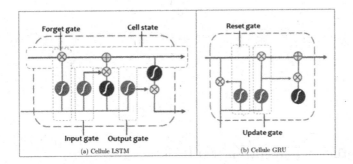

Fig. 2. The architecture of an LSTM and GRU model

3.5 Autoencoder (AE)

An autoencoder is an artificial neural network that learns to reconstruct its original input to output with minimal error. In its simplest form, it consists of an input layer, a hidden layer, and an output layer [20].

The input layer contains the same number of neurons as the output layer. To reduce the dimensions, the hidden layer must be smaller. The figure below shows the architecture of a simple auto-encoder.

The transition from input layer to hidden layer is called encoding stage and the transition from hidden layer to output layer is called decoding stage. We can define these two transitions as:

$$\phi : \chi \rightarrow Z \text{ tel que } x \rightarrow \phi(x) = \sigma(Wx + b) = z \tag{6}$$

$$\psi : Z \rightarrow \chi \text{ tel que } z \rightarrow \psi(z) = \sigma(\hat{W}z + \hat{b}) = \hat{x} \tag{7}$$

with χ and Z respectively the encoder input and output sets, σ, a activation function, W and $\hat{W}$ of the neural network weight matrices and finally, b and $\hat{b}$ bias vectors.

The input vector x is first projected onto the code layer z, called the latent layer. The hidden representation z is then mapped onto the output x using the decoding network. The training therefore consists in minimizing the loss function (Fig. 3):

$$\iota(x, \hat{x}) = || x - \hat{x} ||^2 \tag{8}$$

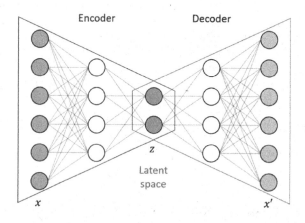

Fig. 3. Architecture of an autoencoder

3.6   Multilayer Perceptron (MLP)

The multilayer perceptron is the most widely used model for neural network applications using the backpropagation training algorithm. The definition of the architecture in an MLP network is a very relevant point, since lack of connections can lead to a network that cannot account for insufficient parameter tuning, while too many connections can lead to overfitting of the data. Learn [21].

The multilayer perceptron is a variant of the original perceptron model proposed by Rosenblatt in [22] in the 1950s. It hides one or more layers between its input and output layers, neurons are organized in layers, connections are always routed from lower layers to upper layers, and neurons in the same layer are not connected to each other (Fig. 4).

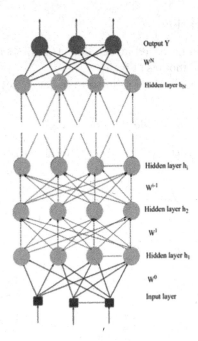

Fig. 4. Feed forward neural network structure

The number of neurons in the input layer corresponds to the number of measures of the pattern problem, and the number of neurons in the output layer corresponds to the number of classes. The choice of the number of layers and the neurons and connections in each layer is called the architecture. Our main goal is to optimize it for a suitable network with enough parameters and good generalization ability for classification or regression tasks.

Learning in an MLP is the process of adjusting link weights to minimize the difference between the output of the network and the desired output. For this, some algorithms such as ant colony [23] are used in the literature. But the most commonly used is called backpropagation, which is based on the gradient descent technique [24].

4   Experiment Results and Discussion

After a detailed study of IDS and the different deep learning methods, and after explaining in detail the design of our choice for the realization of a deep learning based approach, we finally arrived at the final phase of our work which consists in implementing the different phases of our architecture and all the modules and libraries necessary for the proper functioning of our model, and finally, the results obtained by applying the methods already mentioned.

4.1   Binary Classification

We tested our deep learning-based attack detection models with the data set to ensure the fidelity of our models (Fig. 5).

This Table 2 compare the models in terms of accuracy.

Table 2. Performances

The model name	Precision	Recall	F1 Score
MLP	97.24%	98.75%	97.99%.
LSTM	97.77%	97.77%	97.77%
AE	81.49%	98.78%	89.31%

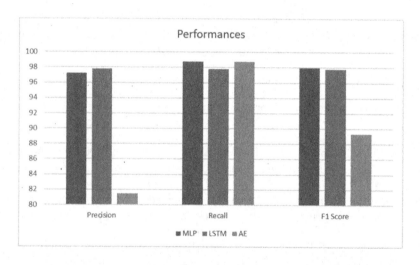

Fig. 5. Performances

The figure below shows the results of the model evaluation (Fig. 6) (Table 3).

Table 3. Training performance on the dataset

The model name	Training Accuracy	Training Loss
MLP	97.85%	0.06%
LSTM	97.77%	0.06%
AE	87.40%	0.76%

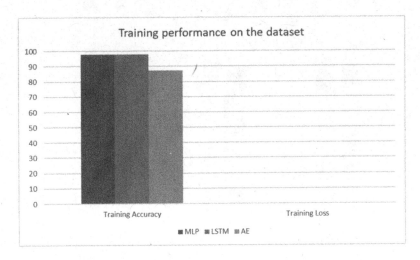

Fig. 6. Training performance on the dataset

In terms of predictive performance, we observed that The best performing model in binary classification is the MLP with an accuracy score equal to 97.85% and a loss of 0.06%.

The results of the other models also remain satisfactory.

4.2   Multiclass Classification

We tested our deep learning based detection model with the dataset and performance metrics listed in the table.

This Table 4 compare the models in terms of accuracy (Fig. 7).

Table 4. Performances

The model name	Precision	Recall	F1 Score
MLP	97.13%	96.81%	96.97%
LSTM	98.04%	97.89%	97.97%
AE	76.95%	88.29%	82.23%

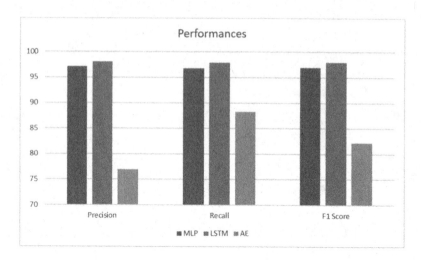

Fig. 7. Performances

The figure below shows the results of the model evaluation (Table 5).

Table 5. Multi-class training performance on the dataset

The model name	Training Accuracy	Training Loss
MLP	96.97%	0.09%
LSTM	97.98%	0.06%
AE	84.41%	0.06%

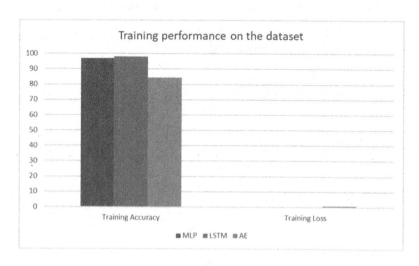

Fig. 8. Performances

In terms of predictive performance, we observed that The most efficient model in munti class classification is the LSTM model with an accuracy score equal to 97.98% and a loss of 0.06%.

The results of the other models also remain satisfactory (Fig. 8).

Conclusion

Cybersecurity encompasses a range of practices aimed at safeguarding vulnerable elements using information and communication technologies (ICT). As part of these practices, intrusion detection systems play a crucial role in complementing existing security modules such as antivirus software and firewalls, which often fall short in countering evolving and sophisticated threats. This study seeks to demonstrate the effectiveness of deep learning techniques in the field of cybersecurity, specifically focusing on intrusion detection.

Our objective is to implement deep learning-based intrusion detection methods and evaluate their performance. To begin, we carefully selected the NSL-KDD dataset, which is widely used for detecting denial of service attacks, probes, remote-to-user, and user-to-root attacks. These types of network cyber-attacks are prevalent and pervasive, as they can be launched remotely and concealed by legitimate user activities on networks. Detecting and preventing such attacks poses significant challenges.

Our primary focus is to explore the detection of these attacks, particularly those that have emerged in recent years. To achieve this, we have implemented two deep learning models: Long Short-Term Memory (LSTM) for supervised learning and autoencoder (AE) for unsupervised learning. Additionally, we have employed a Multilayer Perceptron (MLP) model, a type of artificial neural network, for classification purposes. The results obtained from these models have been highly satisfactory, demonstrating their effectiveness in detecting and mitigating cyber threats.

By leveraging the power of deep learning techniques, we aim to enhance intrusion detection capabilities and contribute to the advancement of cybersecurity practices. Through this study, we strive to provide valuable insights and practical solutions to address the ever-evolving landscape of cyber threats.

Although the initial objectives have been achieved, there are still perspectives and possible improvements to consider for the future. Here are some suggestions:

Exploration of different deep learning methods: It would be interesting to explore other deep learning techniques and conduct a comprehensive comparative study in terms of results, performance, and speed. For example, the use of convolutional neural networks (CNNs) or generative adversarial networks (GANs) can be considered to further enhance the accuracy and robustness of intrusion detection models.

Development of a real-time detection model: A promising perspective would be to create a model capable of capturing and analyzing network traffic in real-time, without relying on a static dataset such as NSL-KDD. This would enable proactive detection of attacks and a faster response to emerging threats. Techniques such as streaming data processing and incremental learning would need to be explored to establish a real-time detection system.

Increasing the diversity of training data: To further improve the performance of detection models, it would be beneficial to expand the diversity of training data. This could involve including more recent attack scenarios, integrating data from different sources or environments, or even collecting domain-specific or industry-specific data.

Optimization of hyperparameters: Optimizing the hyperparameters of detection models can lead to significant improvements in performance. It would be interesting to explore automatic optimization techniques such as Bayesian optimization or random search to find the optimal combinations of hyperparameters for each model.

By pursuing these perspectives and continuing to innovate in the field of deep learning-based intrusion detection, it is possible to further strengthen the ability to effectively and proactively identify and counter cyber threats.

References

1. Lu, Y., Da Xu, L.: Internet of things (iot) cybersecurity research: a review of current research topics. IEEE Internet Things J. **6**(2), 2103–2115 (2018)
2. Al-Yaseen, W.L., Othman, Z.A., Nazri, M.Z.A.: Multi-level hybrid support vector machine and extreme learning machine based on modified k-means for intrusion detection system. Expert Syst. Appl. **67**, 296–303 (2017)
3. Huda, S., Abawajy, J., Alazab, M., Abdollalihian, M., Islam, R., Yearwood, J.: Hybrids of support vector machine wrapper and filter based framework for malware detection. Futur. Gener. Comput. Syst. **55**, 376–390 (2016)
4. Zarpelão, B.B., Miani, R.S., Kawakani, C.T., de Alvarenga, S.C.: A survey of intrusion detection in internet of things. J. Netw. Comput. Appl. **84**, 25–37 (2017)
5. Mienye, I.D., Sun, Y., Wang, Z.: Prediction performance of improved decision tree-based algorithms: a review. Procedia Manuf. **35**, 698–703 (2019)
6. Chowdhury, A., Karmakar, G., Kamruzzaman, J.: The co-evolution of cloud and iot applications: recent and future trends. In: Handbook of Research on the IoT, Cloud Computing, and Wireless Network Optimization, pp. 213–234. IGI Global (2019)
7. Khraisat, A., Gondal, I., Vamplew, P., Kamruzzaman, J.: Survey of intrusion detection systems: techniques, datasets and challenges. Cybersecurity **2**(1), 1–22 (2019). https://doi.org/10.1186/s42400-019-0038-7
8. Javaid, A., Niyaz, Q., Sun, W., Alam, M.: A deep learning approach for network intrusion detection system. Eai Endorsed Trans. Secur. Saf. **3**(9), e2 (2016)
9. Chandola, V., Banerjee, A., Kumar, V.: Anomaly detection: a survey. ACM Comput. Surv. (CSUR) **41**(3), 1–58 (2009)
10. Ertoz, L., et al.: Minds-minnesota intrusion detection system. Next generation data mining, pp. 199–218 (2004)
11. Eskinand, E., Stolfo, S.: Modeling system call for intrusion detection using dynamic window sizes. In: Proceedings of DARPA Information Survivability Conference and Exposition (2001)

12. Qin, M., Hwang, K.: Frequent episode rules for intrusive anomaly detection with internet datamining. In: USENIX Security Symposium, pp. 1–15, Citeseer (2004)
13. Barbará, D., Couto, J., Jajodia, S., Wu, N.: Adam: a testbed for exploring the use of data mining in intrusion detection. ACM SIGMOD Rec. **30**(4), 15–24 (2001)
14. Deshmukh, D.H., Ghorpade, T., Padiya, P.: Intrusion detection system by improved preprocessing methods and naïve bayes classifier using nsl-kdd 99 dataset. In: 2014 International Conference on Electronics and Communication Systems (ICECS), pp. 1–7. IEEE (2014)
15. Yusuf, I.M.: Accuracy of Machine Learning Algorithms in Detecting DoS/DDoS Attacks Types. Ph.D. thesis (2016)
16. Jabbar, M., Samreen, S.: Intelligent network intrusion detection using alternating decision trees. In: 2016 International Conference on Circuits, Controls, Communications and Computing (I4C), pp. 1–6. IEEE (2016)
17. Tavallaee, M., Bagheri, E., Lu, W., Ghorbani, A.A.: A detailed analysis of the kdd cup 99 data set. In: 2009 IEEE Symposium on Computational Intelligence for Security and Defense Applications, pp. 1–6, Ieee, 2009
18. Standardscaler algorithm. https://www.datacorner.fr/feature-scaling. Accessed 28 Apr 2023
19. Yu, Y., Si, X., Hu, C., Zhang, J.: A review of recurrent neural networks: Lstm cells and network architectures. Neural Comput. **31**(7), 1235–1270 (2019)
20. Tschannen, M., Bachem, O., Lucic, M.: Recent advances in autoencoder-based representation learning. arXiv preprint arXiv:1812.05069 (2018)
21. Lins, A., Ludermir, T.B.: Hybrid optimization algorithm for the definition of mlp neural network architectures and weights. In: Fifth International Conference on Hybrid Intelligent Systems (HIS'05), pp. 6-pp. IEEE (2005)
22. Amakdouf, H., El Mallahi, M., Zouhri, A., Tahiri, A., Qjidaa, H.: Classification and recognition of 3d image of charlier moments using a multilayer perceptron architecture. Procedia Comput. Sci. **127**, 226–235 (2018)
23. Socha, K., Blum, C.: An ant colony optimization algorithm for continuous optimization: application to feed-forward neural network training. Neural Comput. Appl. **16**(3), 235–247 (2007)
24. Ramchoun, H., Ghanou, Y., Ettaouil, M., Janati Idrissi, M.À.: Multilayer perceptron: architecture optimization and training (2016)

Diabetes Prediction by Machine Learning Algorithms and Risks Factors

Youssef Fakir[⊠] ⓘ

Sultan Moulay Slimane University, Beni Mellal, Morocco
Youssef.fakir@usms.ma

Abstract. Diabetes is a chronic disease that can have a serious impact on one's health; moreover, the risk of getting it can be decreased with early detection and care. For predicting diabetes, this study aims to compare the performance of six algorithms which are artificial neural networks (ANNs), decision tree (DT), support vector machine (SVM), K-Nearest Neighbors (K-NN), Naive Bayes (NB) and Random Forests models using common risk factors. These models are evaluated in terms of their accuracy, sensitivity, specificity, precision and F-measure. The algorithms were tested using three processes: three factors (glucose, BMI and age), five factors (glucose, BMI, age, insulin and skin) and for the last process all the patterns were used. The variables having the greatest impact on diabetic patients are identified from the association rules extracted, after the extraction of frequent variables by FP-Growth algorithm. By application of the algorithms mentioned above, the results showed that the random forest algorithm is considered as the best machine learning algorithm for the case of all factors but for the cases (3 factors) or (5 factors) Naive Bayes is better compared to the Random Forests algorithm.

Keywords: Artificial Neural Networks · Random Forest · K-nearest Neighbors · Decision Tree · Suport Vector Machine · Naive Bayes · FP-Growth

1 Introduction

In this paper, we are interested in the use of machine learning algorithms for the prediction of type 2 diabetes [1–6], which is a dysfunction of the blood sugar regulation system, in order to reduce the risk of complication of this chronic disease on the patient's health.

Our problem allows us to define medical diagnosis as a classification process and the use of computers is becoming more and more frequent to implement this classification although the doctor's decision is the most important factor in the diagnosis. Classification systems [7–10] are a big help because they reduce errors due to fatigue and the time needed for diagnosis.

This work is characterized by the application of different supervised learning classification algorithms using K nearest neighbors [11–13], Decision Trees [14, 15], Random Forest [16–20], Support Vector Machine [21], Naïve Bayes [11, 16] and ANN [22] to diabetic dataset (PIMA) [23] in order to deduce the best algorithm which will result in an efficient classification of patients in terms of rate of accuracy and sensitivity. Figure 1 illustrates the process of the adopted prediction system.

R. El Ayachi et al. (Eds.): CBI 2023, LNBIP 484, pp. 44–56, 2023.
https://doi.org/10.1007/978-3-031-37872-0_4

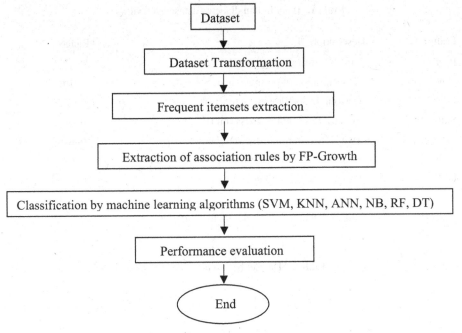

Fig. 1. Prediction system

2   Data Presentation and Analysis

The dataset originally came from the National Institute of Diabetes, Digestive, and Kidney Diseases. The main purpose from this dataset is to realize a diagnostic prediction of whether a patient has diabetes based on some specific diagnostic measures contained in the dataset. Specifically all of the patients here were women who were at least 21 years old and of Pima Indian heritage.

The datasets contains several medical predictor variables and one target variable, Outcome (Table 1). Predictor variables includes the number of pregnancies the patient has had, their BMI, insulin level, age, BloodPressure (BP), SkinThicknes (Skin), BMI, DiabetesPedigreeFunction (DPF), age and outcome [13]. A part of dataset before transformation is given in Table 2. Figure 2 shows the histogram of features. The graph in Fig. 2 shows that the data is unbalanced.

Table 1. Pima Indian diabetes dataset overview

Feature	Description	Range
Preg	-Number of times pregnant	[0–17]
Glucose	-Concentration of plasma glucose	[0–199]
BP	-Dastolic Blood Pressure (mmHg)	[0–122]
Skin	-Triceps skin fold thickness (mm)	[0–99]
Insulin	-Two hour serum Insuline (mu U/ml)	[0–846]
BMI	-Body Mass Index (weight in Kg/(Height in m^2)	[0–67]
DPF	-Diabetes Peggigree Function	[0–2.45]
Age	-Age (years)	[21–81]
Outcome	-Binary value indicating non-diabetic/diabetic	(0,1)

Table 2. Dataset before transformation

	Pregnancie	Glucose	BloodPress	SkinThickn	Insulin	BMI	DiabetesPe	Age	Outcome
1									
2	6	148	72	35	0	33.6	0.627	50	1
3	1	85	66	29	0	26.6	0.351	31	0
4	8	183	64	0	0	23.3	0.672	32	1
5	1	89	66	23	94	28.1	0.167	21	0
6	0	137	40	35	168	43.1	2.288	33	1
7	5	116	74	0	0	25.6	0.201	30	0
8	3	78	50	32	88	31	0.248	26	1

A good dataset is one in which features are strongly correlated to the target class and are strongly uncorrelated to each other. To find the uncorrelated attributes, feature selection is done through a correlation-based approach using a correlation coefficient, which is a number that indicates the strength of the relationship between two variables. There are several types of correlation coefficients, but the most common of all is the Pearson coefficient denoted r, defined by:

$$r = Cov(X, Y)/\sigma_X \sigma_Y \quad (1)$$

where: Cov(X, Y) denotes the covariance of the variables X and Y, σ_X and σ_Y denote their standard deviations. The value of the correlation coefficient between -1 and $+1$.

→ 1 means that they are strongly correlated (strong positive relationship).

→ 0 means no correlation.

→ -1 means that there is a negative correlation (strong negative relationship).

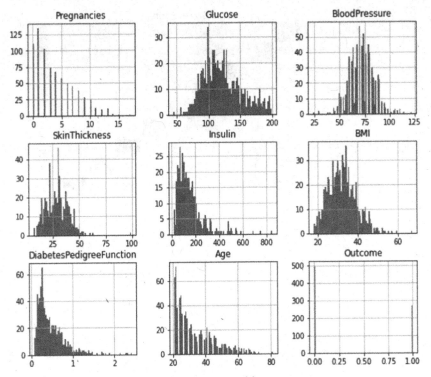

Fig. 2. Histogram of features

Figure 3 illustrates the correlation between the different variables. In the heat map, brighter colors indicate greater correlation. As we can see in Fig. 3, glucose levels, age, BMI and number of pregnancies all have a significant correlation with the outcome variable. Thanks to the correlation analysis result, the attributes Glucose, BMI and Age have a relationship of 0.47, 0.29, 0.24 with Output. By analyzing the values of correlation analysis, glucose, BMI and age have the greatest influence on diabetes in the correlation between variables.

By applying Principal Component Analysis (PCA) [26, 27] to the data (Fig. 4), feature groupings have been observed in the plot above, where arrows that are close to each other represent closely related features. We can see that the following elements are closely related:

– Pregnancy and age
– Blood sugar and blood pressure
– BMI, DPF, insulin level and skin thickness

By analyzing the data (the average of the characteristics), we can see that the characteristics Glucose, BMI, Age, Insulin are very important to help classify the data. On the other hand, PDF, blood pressure and number of pregnancies are very low. These features can be used by machine learning (ML) classification algorithms. This result is provide by FP-Growth in extracting association rules [25].

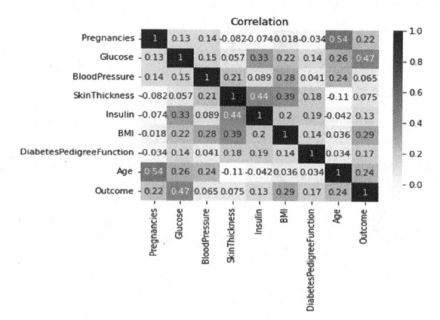

Fig. 3. Heatmap

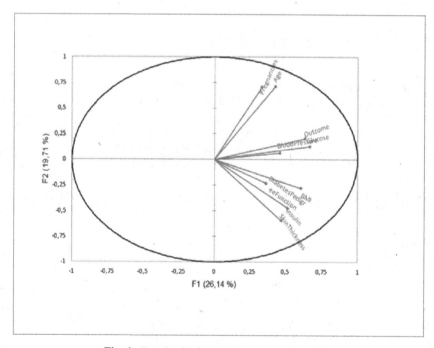

Fig. 4. Result of Principal Component Analysis

3   Evaluation Criteria and Performance Measurement

The endpoint is a key factor both in evaluating the performance of classification and classifier modeling guidance. To compare the performance of the different methods and tools used to our study we use the confusion matrix given in Table 3 where TP, TN, FP and FN denote true positive, true negative, false positive and false negative respectively.

The performance measures used are: Accuracy, Precision, Recall, Specificity, and F-meaure [28].

Table 3. Confusion matrix

	Diabetic (1)	No-Diabetic (0)
Diabetic (1)	VP	FP
No-Diabetic (0)	FN	VN

Accuracy: To predict the performance of a classifier on unobserved data, we use the evaluation method called Accuracy, it is defined as the ratio between the well-classified examples and the total number of examples:

$$Accuracy = \frac{TP + FN}{TP + TN + FP + FN} \tag{2}$$

Precision: precision is the proportion of well-classified items for a given class.

$$Precision = \frac{TP}{TP + FP} \tag{3}$$

Recall (Sensitivity): Sensitivity is used to represent the frequency of positive responses to the test. In other words, it is the conditional probability of having a positive test result if M (disease) is present.

$$Sensitivity = \frac{TP}{TP + FN} \tag{4}$$

Specificity: Specificity represents the frequency of negative test responses among patients who do not have diabete disease. In addition, it is the probability conditional on having a negative test result with the absence of diabete disease. It corresponds to 1 if there is no false positive.

$$Specificity = \frac{TN}{TN + FP} \tag{5}$$

F-measure: Represents the weighted average of Precision and Sensitivity. These last two make a fair relative contribution to the F-measure.

$$F\text{-}measure = \frac{2 * Sensitivity * precision}{Sensitivity + precision} \tag{6}$$

4 Experimental Results

This part relates the results of the comparative study between the classification tools data using the six algorithms (RF, DT, ANN, SVM, KNN, NB) applied to the data base (PIMA). The objective of each algorithm is to achieve a better classification of future observations by minimizing classification error. The algorithms were tested using three process: three factors (glucose, BMI and age), five factors (glucose, BMI, age, insulin and skin) and for the last process we used all the patterns. The five models were evaluated based on accuracy, sentivity, specificity, precision and F-measure. The detailed predictions are presented in the form of confusion matrices. The tables below show the confusion matrices for each algorithm used.

4.1 Prediction Using All the Factors in the Database

The tables below show respectively the confusion matrices obtained by using the six classification techniques (Neural Network, Support Vector Machine, Decision Tree, Random Forest, Naive Bayes and k-Nearest Neighbor). These algorithms were applied to the training set to predict the results of the test set which yielded the result shown in Table 10. One observation from the table is that the accuracy of all methods is more on our dataset since the former has more relevant fields to assess diabetes risk. The Random Forest model has the best score in terms of accuracy and precision (Tables 4, 5, 6, 7, 8 and 9).

Table 4. KNN confusion matrix

	Diabetic	No_diabetic
Diabetic	VP = 134	FN = 16
No_diabetic	FP = 28	VN = 53

Table 5. RF confusion matrix.

	Diabetic	No_diabetic
Diabetic	VP = 139	FN = 11
No_diabetic	FP = 16	VN = 65

Table 6. SVM confusion matrix.

	Diabetic	No_diabetic
Diabetic	VP = 125	FN = 25
No_diabetic	FP = 29	VN = 52

Table 7. ANN confusion matrix.

	Diabetic	No_diabetic
Diabetic	VP = 134	FN = 16
No_diabetic	FP = 16	VN = 65

Table 8. DT confusion matrix.

	Diabetic	No_diabetic
Diabetic	VP = 133	FN = 17
No_diabetic	FP = 21	VN = 60

Table 9. NB confusion matrix.

	Diabetic	No_diabetic
Diabetic	VP = 130	FN = 31
No_diabetic	FP = 21	VN = 49

Table 10. Metric for different algorithms

Algorithms	Accuracy (%)	Precision (%)	Sensitivity (%)	Specificity (%)	F-measure (%)
KNN	81	80	77	65,43	78
RF	**88,31**	**88**	**86**	**80,24**	**87**
SVM	77	74	74	64,19	74
ANN	**86**	**85**	**85**	**80,24**	**85**
NB	77,48	80,74	**86,09**	**74**	**87,86**
DT	84	82	81	70	82

We can say that the Random Forest model is the one that has the best performance in general, taking into account the fact that its scores for the four metrics are more or less close to each other with the best possible recall, i.e. -say a better ability to find all the positive instances (of the sick class which is the class of interest). From Table 10 and Fig. 5 which shows the graphical performance of all classification algorithms based on ranked instances we can conclude that the Random Forest classification algorithm comparatively outperforms other algorithms.

The Random Forest algorithm is considered the best supervised machine learning method of this experiment because it gives higher accuracy compared to other classification algorithms with an accuracy of 88.31%.

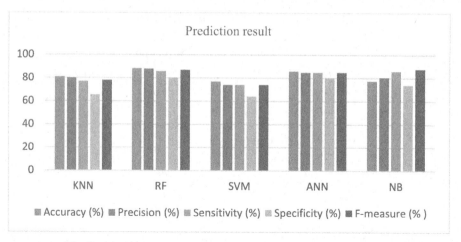

Fig. 5. Comparison of results with different metrics (with all factors)

4.2 Prediction by Three Factors

The attributes (items) are glucose, BMI, and age. The confusion matrices using the Random forest and Naive Bayes algorithms are shown in Table 11 and Table 12 respectively. The values of precision, accuracy, specificity, recall and F measure are given in Table 13 and illustrated in Fig. 6. The Random Forest algorithm is considered the best-supervised machine learning method of this experiment.

Table 11. RF confusion matrix (three factors)

	Diabetic	No_diabetic
Diabetic	VP = 123	FN = 28
No_diabetic	FP = 31	VN = 48

Table 12. NB confusion matrix.

	Diabetic	No_diabetic
Diabetic	VP = 133	FN = 30
No_diabetic	FP = 18	VN = 49

Table 13. Result for three factors

Algorithms	Accuracy (%)	Precision (%)	Sensitivity (%)	Specificity (%)	F-measure (%)
RF	**75,22**	**82,12**	**80,52**	**64,47**	**81,31**
NB	79,13	81,60	**88,08**	**62,03**	**84,71**

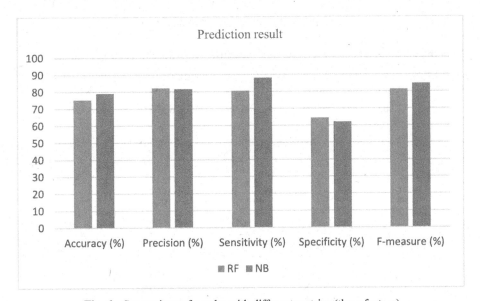

Fig. 6. Comparison of results with different metrics (three factors)

4.3 Prediction by Five Factors

The factors used are glucose, BMI, age, insulin and skin. The confusion matrices using the Random forest and Naive Bayes algorithms are shown in Table 14 and Table 15 respectively. The values of precision, accuracy, specificity, recall and F-measure are given in Table 16 and illustrated in Fig. 7. The Random Forest algorithm is considered the best supervised machine learning method of this experiment.

Table 14. RF confusion matrix.

	Diabetic	No_diabetic
Diabetic	VP = 121	FN = 30
No_diabetic	FP = 30	VN = 49

Table 15. NB confusion matrix

	Diabetic	No_diabetic
Diabetic	VP = 130	FN = 30
No_diabetic	FP = 21	VN = 49

Table 16. Result by using five factors

Algorithms	Accuracy (%)	Precision (%) (%)	Sensitivity (% (%)	Specificity (%) (%)	F-measure (%)
RF	**73,91**	**80,79**	**79,74**	**62,34**	**80,26**
NB	77,83	81,25	**86,09**	**62,03**	**83,60**

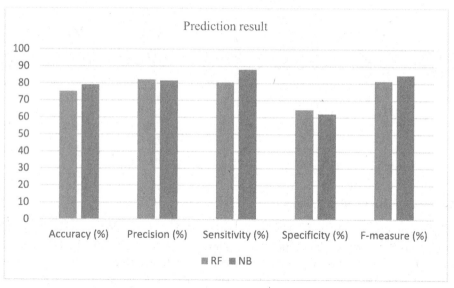

Fig. 7. Comparaison result for different metrics (five factors).

5 Conclusion and Futurs Works

This work led us to the development of a comparative study between six algorithms which are ANN, DT, RF, SVM and KNN applied to a selected database of the medical field. The objective is to give professionals the opportunity to study and apply the different classification techniques and tools. By application of the algorithms mentioned above, the results showed that the random forest algorithm is considered as the best machine learning algorithm for the case of all patterns but for the cases (three factors) or (five factors) Naive Bayes is better compared to the Random Forests algorithm.

This work need to be mproved using other methods such as deep learning and tranformer neural network.

References

1. Larabi-Marie-Sainte, S., Aburahmah, L., Almohaini, R., Saba, T.: Current techniques for diabetes prediction: review and case study. Appl. Sci. **9**(21), 4604 (2019). https://doi.org/10.3390/app9214604
2. Divya, K., Sirohi, A., Pande, S., Malik, R.: An IoMT assisted heart disease diagnostic system using machine learning techniques. In: Hassanien, A.E., Khamparia, A., Gupta, D., Shankar, K., Slowik, A., (eds.) Cognitive Internet of Medical Things for smart healthcare, pp. 145–161. Springer, New York (2021). https://doi.org/10.1007/978-3-030-55833-8_9
3. Kumar, P.M., Devi, G.U.: A novel three-tier Internet of Things architecture with machine learning algorithm for early detection of heart diseases. Comput. Electr. Eng. **65**, 222–235 (2018). https://doi.org/10.1016/j.compeleceng.2017.09.001
4. Komi, M., Li, J., Zhai, Y., Zhang, X.:. Application of data mining methods in diabetes prediction. In: 2017 2nd International Conference on Image, Vision and Computing (ICIVC), Chengdu, China, pp. 1006–1010 (2017). https://doi.org/10.1109/ICIVC.2017.7984706
5. Samant, P., Agarwal, R.: Machine learning techniques for medical diagnosis of diabetes using iris images. Comput. Methods Prog. Biomed. **157**, 121–128 (2018). https://doi.org/10.1016/j.cmpb.2018.01.004
6. Samant, P., Agarwal, R.: Comparative analysis of classification based algorithms for diabetes diagnosis using iris images. J. Med. Eng. Technol. **42**, 35–42 (2018). https://doi.org/10.1080/03091902.2017.1412521
7. You, J., van der Klein, S.A.S., Lou, E., Zuidhof, M.J.: Application of random forest classification to predict daily oviposition events in broiler breeders fed by precision feeding system. Comput. Electron. Agric. **175**, 105526 (2020). https://doi.org/10.1016/j.compag.2020.105526
8. Burdi, F., Setianingrum, A.H., Hakiem, N.: Application of the Naive Bayes method to a decision support system to provide discounts (case study: PT. Bina Usaha Teknik). In: 2016 6th International Conference on Information and Communication Technology for The Muslim World (ICT4M). Jakarta, pp. 281–285 (2016). https://doi.org/10.1109/ICT4M.2016.064
9. Akbar, R., Nasution, S.M., Prasasti, A.L.: Implementation of Naive Bayes algorithm on IoT-based smart laundry mobile application system. In: 2020 international conference on information technology systems and innovation (ICITSI). Bandung - Padang, Indonesia, pp. 8–13 (2020). https://doi.org/10.1109/ICITSI50517.2020.9264938
10. Pandiangan, N., Buono, M.L.C., Loppies, S.H.D.: Implementation of decision tree and Naïve Bayes classification method for predicting study period. J. Phys. Conf. Ser. **1569**, 022022 (2020). https://doi.org/10.1088/1742-6596/1569/2/022022
11. Gomathi, S., Narayani, V.: Monitoring of lupus disease using decision tree induction classification algorithm. In: 2015 International Conference on Advanced Computing and Communication Systems. Coimbatore, India, pp. 1–6 (2015). https://doi.org/10.1109/ICACCS.2015.7324054
12. Abdar, M., Nasarian, E., Zhou, X., Bargshady, G., Wijayaningrum, V.N., Hussain, S.: Performance improvement of decision trees for diagnosis of coronary artery disease using multi filtering approach. In: 2019 IEEE 4th International Conference on Computer and Communication Systems (ICCCS). Singapore, pp. 26–30 (2019). https://doi.org/10.1109/CCOMS.2019.8821633
13. Premamayudu, B., et al.: Diabetes prediction using machine learning KNN -algorithm technique. Int. J. Innovative Science Res. Technol. **7**(5) (2022)

14. Jadhav, S.D., Channe, H.P.: Comparative study of K-NN, naive bayes and decision tree classification techniques. Int. J. Sci. Res. **5**(1), 1842–1845 (2016)
15. Wu, X., Wang, S., Zhang, Y.: Review of K nearest neighbor algorithm theory and application. Comput. Eng. Appl. **53**(21), 1–7 (2017)
16. Kuswanto, H., Mubarok, R.: Classification of cancer drug compounds for radiation protection optimization using CART. In : The Fifth Information Systems International Conference (2019)
17. Shirole, U., Joshi, M., Bagul, P. : Cardiac, diabetic and normal subjects classification using decision tree and result confirmation through orthostatic stress index. Informatics in Medicine Unlocked **17**, 100252 (2019)
18. Xu, W., Jiang, L.: An attribute value frequency-based instance weighting filter for naive Bayes. J. Exp. Theor. Artif. Intell. **31**(4), 225–236 (2019)
19. Svetnik, V., Liaw, A., Tong, C., Culberson, J.C., Sheridan, R.F., Feuston, B.P.: Random forest: a classification and regression tool for compound classification and QSAR modeling. J. Chem. Inf. Comput. Sci. **43**(6), 1947–1958 (2003)
20. Matsumoto, A., Aoki, S., Ohwada, H.: Comparison of random forest and SVM for raw data in drug discovery: prediction of radiation protection and toxicity case study. Int. J. Machine Learning Comput. **6**(2), 145–148 (2016)
21. Zekić-Sušaca, M., Hasa, A., Knežev, M.: Predicting energy cost of public buildings by artificial neural networks, CART, and random forest Forest. Neurocomputing **439**, 223-233 (2021)
22. Butwall, M., Kumar, S. : A data mining approach for the diagnosis of diabetes mellitus using random forest classifier. Int. J. Computer Appl. **120**(8) (2015)
23. Kuswanto, H., Mubarok, R., Ohwada, H.: Classification using naive bayes to predict radiation protection in cancer drug discovery: a case of mixture based grouped data. Int. J. Artificial Intell. **17**(1), 186–203 (2019)
24. Wadiai, Y., Baslam, M.: Machine learning approach to automate decision support on information system attacks. Lecture Notes in Business Information Processing ISBN 978–3–031–06457–9 ISBN 978–3–031–06458–6 (eBook) https://doi.org/10.1007/978-3-031-064 58-6
25. Fakir, Y., Maarouf, A., El Ayachi, R.: Mining frequents itemset and association rules in diabetic dataset. Lecture Notes in Business Information Processing ISBN 978–3–031–06457–9 ISBN 978–3–031–06458–6 (eBook) https://doi.org/10.1007/978-3-031-06458-6
26. Bair, E., Hastie, T., Paul, D., Tibshirani, R. : Prediction by supervised principal components. J. American Statistical Assoc. **101**(473), 119–137 (2006)
27. Borges, V.R.P., Esteves, S.L., De Nardi Araujo, P., Oliveira, L.C., Holanda, M. : Using Principal Component Analysis to support students' performance prediction and data analysis, VII Congresso Brasileiro de Informática na Educação (CBIE 2018), Anais do XXIX Simpósio Brasileiro de Informática na Educação (SBIE 2018)
28. Fakir, Y., Abdelmotalib, N. : Analysis of decision tree algorithms for diabetes prediction. Lecture Notes in Business Information Processing ISBN 978–3–031–06457–9 ISBN 978–3–031–06458–6 (eBook) https://doi.org/10.1007/978-3-031-06458-6

Machine Learning Prediction of Weather-Induced Road Crash Events for Experienced and Novice Drivers: Insights from a Driving Simulator Study

Zouhair Elamrani Abou Elassad[1]([⊠]), Mohamed Ameksa[2],
Dauha Elamrani Abou Elassad[1], and Hajar Mousannif[2]

[1] SARS Research Team, ENSAS, Cadi Ayyad University, Safi, Morocco
z.elamrani@uca.ma
[2] LISI Laboratory, FSSM, Cadi Ayyad University, Marrakesh, Morocco

Abstract. Road crashes are one of the most critical issues that pose a serious threat to our daily life; Crash occurrences prediction is a key role in designing efficient intelligent transportation systems. In this study, we aim to analyze road crash events for experienced and novice drivers under several weather conditions during multiple driving simulations that have been conducted using a desktop driving simulator. This work outlined the effect of snow and rain conditions on driver behavior by endorsing real-time driver data namely: wheel angle position, throttle pedal position and brake pedal position. Moreover, optimized modeling strategies using the deep learning algorithm Multilayer Perceptron (MLP) and Support Vector Machine (SVM) along with Bayesian Networks (BN) models have been developed to analyze crash events. To the authors' knowledge, there has been a limited interest at assessing the impact of both snow and rainy weather conditions on the occurrence of crash events while providing a critical analysis for experienced and novice drivers based on driver entries; this approach fill the research gap of the combined effect of driving experience and weather conditions on road crash occurrence. The findings depict superior performances have been obtained when adopting the proposed strategy. As a whole, new insights into weather-induced crash events' investigation for experienced and novice drivers have been acquired and can be endorsed for designing effective crash avoidance/warning systems.

Keywords: Crash prediction · Machine learning · Novice drivers · Experienced drivers · Weather conditions · Data balancing · Driving simulator

1 Introduction

Nowadays, traffic accidents have become a paramount concern and a pressing issue in society, causing numerous health complications, financial setbacks, and loss of lives. According to a report by the World Health Organization [1], approximately 1.35 million individuals succumb to road traffic collisions annually, with an additional 20–50 million people worldwide suffering injuries or disabilities. Traditionally, adhering to safe driving practices necessitates drivers' ability to promptly and reliably react to hazards encountered on the road. In fact, studies have revealed that inexperienced drivers, particularly those who are new to the driving experience, exhibit a higher incidence of accidents compared to individuals in other age groups when engaging in motorized activities [2, 3].

Numerous research studies have affirmed that the impact of weather events on traffic maneuvers and safety outcomes has emerged as a significant concern, and the reduction in visibility caused by rain is a grave matter. Driving in rainy conditions is deemed as an exceedingly hazardous occurrence, as drivers tend to modify their driving behavior to accommodate the challenges posed by inclement weather. Depending on the prevailing environmental conditions, drivers exercise increased vigilance by maintaining greater distances between vehicles and reducing their driving speeds. Although most vehicles decelerate during rainfall due to impaired vision and decreased surface traction, accidents still transpire [4]. In an effort to examine road accidents and devise effective traffic management strategies, researchers have increasingly scrutinized risky driving behaviors [5, 6]. The analysis of accidents is a complex process influenced by various factors such as driver conduct, weather conditions, and environmental elements [7]. When studying accidents, the assessment of pre-incident risk exposure is commonly associated with diverse traffic flow measures, which underscore the crucial role of driver inputs in ensuring safe driving and identifying collision incidents. Evaluating driving behavior through the utilization of accelerator and brake pedals, as well as steering wheel inputs, has been a prevalent approach to proactively facilitate potential countermeasures and enhance road safety systems [7, 8]. Consequently, the data required for accident analysis can be obtained through various experimental methods, including naturalistic driving studies, field driving studies, and driving simulator studies [9]. In this study, we have opted to employ simulator experiments, as they have been extensively utilized in transportation research due to their significant advantages of maintaining complete empirical control over conditions and facilitating the examination of multiple design configurations. Moreover, driving simulations conducted in secure environments are particularly valuable for investigating crash incidents.

Crash events are typically investigated through two main approaches: statistical learning-based and machine learning (ML)-based methods. While classical statistical techniques like logistic regression [10] and discriminant analysis [11] which have been widely utilized in crash analysis, they often suffer from data quality issues and rely heavily on extensive historical data [12]. On the other hand, machine learning models have demonstrated superior performance in predicting future events and have shown satisfactory outcomes in various transportation systems [13]. The appeal of ML models lies in several key factors. Firstly, they excel in addressing complex non-linear problems by leveraging datasets from multiple sources. Secondly, they possess the flexibility to incorporate new data to enhance estimation performance. Lastly, they offer predictive and explanatory capabilities through the extraction of rules. Machine learning techniques such as Artificial Neural Network MultiLayer Perceptron (MLP), Support Vector Machines (SVM), and Bayesian Networks (BN) have emerged as popular choices for crash investigation, consistently outperforming other modeling techniques [14–16]. MLP has gained popularity due to its effectiveness in handling complex tasks, learning data representations in supervised and unsupervised settings, and its parallel process-ing capabilities and fault tolerance [17]. SVM has shown promise in evaluating safety performance measures for vehicle crashes, displaying better goodness-of-fit compared to negative binomial models. Additionally, SVM demonstrates proficiency in handling small data sizes, minimizing overfitting, and exhibiting superior generalization abilities [18]. BN, with its stochastic processes between nodes, proves to be a valuable ML tech-nique for risk assessment problems with low probability and high uncertainty [19]. It has been widely and successfully employed in domains with significant uncertainty and has found extensive application in various data mining problems. Although several studies have focused on analyzing crash events under different weather conditions, most of the previous research has relied on crash data derived from police reports, which may be susceptible to inaccuracies. The reported conditions may reflect the observations of the person filling out the crash report rather than the actual weather conditions at the time of the accident [20]. Furthermore, limited effort has been devoted to quantifying the influ-ence of real-time data on rainy and snowy weather conditions, as well as driver inputs, in predicting crash collisions for both novice and experienced drivers. Thus, within this context, the development of a reliable crash forecasting strategy for proactive safety analysis is unquestionably crucial and necessary.

In this present paper, an efficient imbalance-learning approach for predicting road crash events for novice and experienced drivers is elaborated. Various day-time driving simulations have been conducted using a Desktop driving simulator to improve our understanding of the inexperienced driving behavior leading to crash collisions under different weather conditions. Furthermore, this work depicted the effect of both rain and snow on traffic safety by endorsing driver inputs determined as throttle pedal position, brake pedal position and wheel angle collected in real-time across the various trails. Moreover, an imbalance-learning strategy based on SMOTE technique has been endorsed in order to construct effective machine learning techniques based on SVM, MLP and BN machine learning models. There were, consequently, substantial gaps in the literature, as there appeared to be little to no research providing advanced prediction of road crash events for both novice and experienced drivers under rainy and snowy weather patterns while accounting for class-imbalance.

The remainder of this study is organized as follows. Section 2, discusses related research focusing on previous efforts in crash events classification. Section 3, in the methodology section, the development of modeling techniques is presented in details. In Sect. 4, the results are reported and interpreted. Finally, conclusions with future scopes of the present study are offered.

2   Related Work

The investigation of road collisions has revealed that hazardous traffic conditions can be attributed to driver mistakes and environmental factors, including driving behavior and weather conditions [7]. A pertinent research study outlined five distinct phases in the progression of a crash: normal driving, deviation from normal driving, emerging situation, critical situation, and inevitable crash [19]. The study also delved into the causes that lead from one phase to another. Another approach, as presented in the work of [21], aimed to identify and validate near-accidents, accidents, and normal driving behavior using naturalistic driving data. This process involved four primary stages: filtering, classification, validation, and modeling.

Machine learning techniques have proven to be highly valuable in real-time crash analysis, enabling the identification of relationships between accident occurrences and various associated factors or contemporary situations. Support Vector Machines [22], Neural Networks [23], and Bayesian Networks [14, 15] are among the commonly implemented machine learning approaches. Artificial Neural Networks (ANNs) are particularly effective in handling noisy data and performing fast real-time computations with robust efficiency [7]. In terms of driving behavior, inexperienced drivers are less proficient in identifying and responding to hazards compared to more experienced drivers. Young novice drivers possess certain characteristics that increase their vulnerability to injuries or fatalities in traffic crashes. These factors include physiological aspects, underestimation of risks, and underdeveloped hazard awareness skills [24]. Additionally, inexperienced novice drivers often exhibit a higher tendency for sensation seeking, and their driving behavior can be influenced by the presence of their companions [25].

With According to data from the National Highway Traffic Safety Administration, weather conditions contribute to more than 22% of total annual crashes. Specifically, a significant portion of weather-related accidents occurs on wet roads and during rainfall, accounting for approximately 73% and 46% of crashes, respectively [26]. It has been observed that the likelihood of a crash is 70% higher in rainy conditions compared to clear weather [20]. While numerous studies have explored the impact of weather conditions on crash events [27–29], it is important to note that many of these investigations rely on weather variables obtained from police crash reports. However, these reports may be susceptible to inaccuracies as the conditions reported might reflect the observations of the person completing the report rather than the actual weather conditions at the time of the accident [30].

Driving The use of driving simulators has been widely embraced in transportation research due to their ability to provide complete experimental control over scenarios and the opportunity to explore multiple design configurations in a safe environment [9]. While there have been several studies examining the influence of weather conditions on predicting crash incidents, the prediction of crashes across various weather patterns remains relatively unexplored. Additionally, most research that incorporates weather variables in crash analysis relies on data collected from police reports, which may be prone to inaccuracies. Therefore, the utilization of simulator-generated weather patterns offers the advantage of real-time, controllable, and reliable conditions, thereby driving our interest in exploring such features. Building upon the aforementioned analysis, there is a need for a more effective crash assessment that can provide drivers and road managers with dependable information by integrating various computational techniques.

3 Methodology

This work aims to provide an effective crash prediction for novice and experienced drivers based on an imbalance-learning strategy combined with highly efficient machine learning models using driver inputs namely throttle and brake pedals' positions along with the steering angle, and under rainy and snowy weather covariates. This section describes the methodology used to capture various features, employing machine learning classifiers and the adopted data-balancing techniques.

3.1 Driving Simulations

Participants and Apparatus
The study was conducted out using a fixed-based driving simulator located at the University of Cadi Ayyad (UCA) facility. Simulator driving experiments offer a notable benefit by allowing the replication of real-world driving experiences within a safe and controlled environment. These simulations encompass a wide range of driving conditions, including various climates, terrains, and traffic scenarios [9]. Certainly, it would be very risky to carry out trials on real road settings. The driving simulation was run through the Project Cars 2 simulator. In reference to the provided information about the experiment's general intentions, all participants gave informed consent form about data recording of their driving performance.

The driving simulation was run through the Project Cars 2 virtual simulator [31]; a DELL XPS running on Windows 10 and a 2015 MacBook Pro with an i7 2.8 GHz chip, 16GB RAM and SSD hard drive were employed for data collection and computational analyses. Participants viewed the simulation on a 27-inch LCD monitor with a resolution of 1920x1080 pixels and heard auditory via a surround speaker system. The computer was fitted with a Logitech® G27 Racing Wheel set (steering wheel, accelerator pedal, and brake pedal) with the adjustable Logitech Evolution® PlaySeat, simulations were

Fig. 1. The driving simulator setup at UCA

conducted with automatic gear selection, thus gear shifter was not needed. Figure 1 depicts the driving simulator setup.

Simulation Scenario and Data Collection

The drivers were appointed to a quiet laboratory to virtually drive the vehicle; each participant navigated experimental session drives on a virtual two-lane urban road required about 20 min to complete when the speed limits are kept. An overview of the selected layout of the driving route is depicted in Fig. 2.

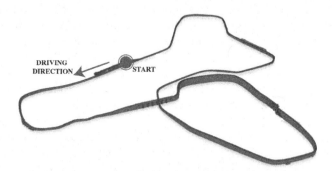

Fig. 2. Scenario roadway layout for driving trials

The driving scenario was performed during day-time and under two different sequential weather patterns, rain which can been seen in Fig. 3.a, and snow that is illustrated in Fig. 3.b; the scenario aimed to simulate various intricacies and aspects that real-world driving entails to collect enough raw data before the crash. The adopted protocol had similar traffic conditions and the identical number of outer events for all participants.

Upon their arrival, participants were required to provide their informed consent by signing a form, indicating their willingness to take part in the experiment. The experimental session involved two visits to the simulator, during which drivers were instructed to drive in a manner consistent with real-world driving, adhering to traffic regulations. The initial visit served as a 20-min trial drive to familiarize the participants with the simulator. In the subsequent visit, which served as the main session, drivers maneuvered the vehicle on a virtual roadway featuring two different weather conditions, alongside typical scenarios encountered on the route, such as other vehicles. Throughout each drive, driving data was continuously recorded at a sampling frequency of 20 Hz using the UDP protocol. The collected data encompassed driver inputs, including throttle/brake pedal position and steering angle, as well as weather information, such as light rain and heavy rain.

The pedal input (Fig. 4a) is determined by the distance between the pedal's current position relative to the neutral position while steering-wheel angle, shown in the right side of Fig. 4, is the angle computed between the actual placement of the vertical center of the steering wheel and the resting location of the wheel. In return, the dependent variable is crash event, coded as a binary variable with a value of 1 if a crash was identified and 0 if not. Apart from the categorical data of weather season, all variables are with continuous.

Fig. 3. Weather patterns in driving scenario: (a) rain, (b) snow

A thorough and comprehensive data screening that includes cleaning and consistency checks is executed to secure data operability and validity for the analysis.

The definition of novice drivers in this paper was consistent with a previous study [25], and experienced drivers were divided into two sets of participants according to their driving experience, as follows:

- Set 1: driving experience ≤ 2 years (i.e. novice driver).
- Set 2: 2 years $<$ driving experience ≤ 10 years (i.e. experienced driver).
- Set 3: 10 years $>$ driving experience (i.e. experienced driver).

The distribution of these sets of subjects was as follows: Set 1 = 42%, Set 2 = 33% and Set 3 = 25%.

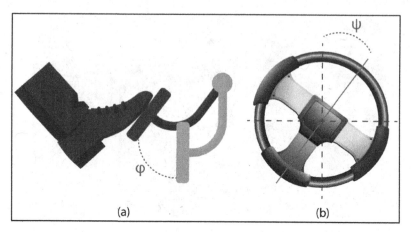

Fig. 4. The driving simulator setup at UCA

3.2 Data Balancing

Crash analysis usually leads to dealing with unbalanced data sets due to unequal outcome classes distribution. This sort of imbalances produce a bias toward the majority class, since modeling classifiers prioritize the class with the higher number of instances (crash instances) leading to an over-prediction of this class [32]. In this work, the synthetic minority over-sampling technique (SMOTE), presented by [33], was applied to address this issue.

SMOTE The SMOTE technique generates synthetic samples for the minority class by considering random intervals between existing minority instances instead of simply duplicating them. Initially, the technique identifies the k-nearest neighbors for each minority case, following the recommendation of [33], with a specified value of k as 5. Subsequently, multiple iterations are performed based on the desired over-sampling, where one neighbor is randomly selected from the k-nearest neighbors. The difference between the current instance and its neighbor is then calculated and multiplied by a random number between 0 and 1. Finally, the newly generated synthetic instances are added to the dataset and assigned to the minority class. Unlike random oversampling techniques that duplicate the original minority instances randomly to rebalance the data, SMOTE achieves balance without data replication, thus mitigating the risk of overfitting [14, 32]. The effectiveness and popularity of SMOTE stem from its computational efficiency, simplicity, and notable performance [34, 35]. Figure 3 provides an illustration of the SMOTE algorithm (Fig. 5).

3.3 Prediction Models

In this study, both MLP, SVM and BN have been applied for the assessment of crash events. MLP are abstract computational techniques inspired by biological neural networks and are efficient and applicable for predicting the relation between dependent and independent parameters. The performance of MLP prediction is highly affected by its structure which is comprised of an input layer, hidden layers and an output layer; each

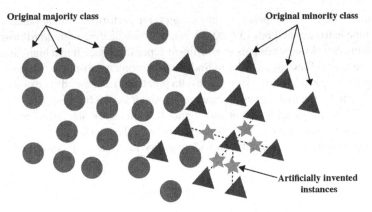

Fig. 5. Overview of the SMOTE procedure.

layer contains a group interconnected processing units called neurons or nodes that are effective in processing enormous parallel computations and knowledge representation [36]. In every layer, Neurons are connected to other neurons in previous and following layers with a connection illustrating the synapse that transmits data from the previous neuron and multiplies it by a specific "weight" based on the information strength in determining the output. To train a neural network, forward propagation and backward propagation are used repeatedly to calibrate all the network weights. First, the forward propagation is used by applying a set of randomly selected weights to calculate the output. Then based on the error between model output and desired output, the backward propagation adjusts the weights to decrease the error and stops when it reaches a specified maximum iteration number or satisfies other stopping criteria. The outputs of an ANN model rely on internal structure, connection weights, activation function and bias value, which can be described as follows:

$$y = f(\sum_j w_{ij}x_j + b) \tag{1}$$

where f is the activation function, w is the weight value, x is the input vector, b and is the bias value. The terminology "Deep Learning" is used to express the model of multiple hidden layers architecture which are depicted to examine a complex problem [37]. The number of hidden layers, neurons in each layer, and activation function were first determined with the aim of minimizing the simulation error in the training phase and were identified by the trial and error strategy, as such, three hidden layers (5, 6, 8 neurons) were employed for the model architecture.

Conversely, SVM, Developed by [38], is a non-probabilistic binary linear classifier that can be used to solve a classification problem by constructing optimal separating hyperplane in a manner that the margin is maximized so that SVM has good generalization ability. A set of related supervised learning methods used for prediction and regression based on the statistical learning theory and structural risk minimization which are the theoretical foundations for the SVMs learning algorithms. It has been proven that SVM is highly efficient and robust algorithm for binary classification problems, and it has been found to demonstrate proportionate or superior performance than other statistical and machine learning methods [39, 40]. Whereas when dealing with non-linear separable problems, SVM maps samples in non-linear separable space into a high dimensional space using kernel function in order to find the separating hyper plane. The kernel functions map the original linearly inseparable data points into a higher dimensional space in which they can be separated linearly. The most popular kernel functions are Radial Basis Function (RBF), polynomial, linear and sigmoid. In this work, the RBF kernel function in Eq. (1) has been adopted as it has been proven to be the best kernel function due to its computational efficiency, veracity, and adaptability in processing complex parameters [41].

$$K(x_i, x_j) = \exp(-\gamma ||x_i - x_j||^2) \tag{2}$$

where $|| x_i - x_j ||$ is the Euclidean distance between vectors of features computed from training or testing data set, and γ is the bandwidth of the Radial Basis Function.

In comparison, Bayesian networks are probabilistic graphical models, which graphically represent the conditional dependence relations between its variables [42]. As any probabilistic model, it has some assumptions, that in this case are related to the joint distribution of the variables in the problem. A Bayesian network falls into the class of Directed Acyclic Graphs (DAGs). This means that, in the graph, the links between the variables have a direction, representing a direct dependency between them. In a directed graph, we define a relation between parent variables and child variables, that is represented by an arrow from a parent to a child. These parent-child relations between the variables can be complex, as long as the graph continues to be directed and acyclic.

The fundamental property of a Bayesian network is that the joint probability distribution of all the variables in the model factorizes such as a given variable only depends directly on its parent nodes. In this way, a Bayesian network can be characterized from all the conditional probability distributions $P(X_i|Pa_i)$, where Pa_i represents all the parent variables of X_i. Thus, the joint probability function of all the variables in the model can be obtained by the product of all the conditional probability functions as:

$$P(X_1, \ldots X_k) = P(X_k|Pa_k) \tag{3}$$

A key advantage of Bayesian networks is their availability to visualize conditional independence relations in the variables, by simply exploring the graph visually.

3.4 Performance Metrics

The accuracy, Precision and Recall performance metrics are used to evaluate the quality of the classification models. The True Positive (TP) indicates the number of crash occurrences correctly classified, and False Positive (FP) indicates the number of non-crash events incorrectly classified as crash-events. False Negative (FN) indicates the number of crash events incorrectly classified as non-crash events, and True Negative (TN) indicates the number of non-crash occurrences correctly classified. As such, accuracy specifies the number of correct predictions made by the model over all kinds predictions made. Recall is defined as the proportion of correctly classified positives (i.e. crash events correctly classified). Since the primary goals focus is to correctly predict the rare events of the accident class, Recall is a particularly substantial metric of classifier performance in this case. Precision on the other hand is a measure of accuracy outlining the relevance ratio of the predicted instances, i.e. percentage of truly predicted events from all predicted events.

$$Accuracy = \frac{TP + TN}{TP + FN + FP + TN} \qquad (4)$$

$$Recall = \frac{TP}{TP + FN} \qquad (5)$$

$$Precision = \frac{TP}{TP + FP} \qquad (6)$$

Data were divided into training and validation sets. For the evaluation of the classification performance of each classifier and with the aim of obtaining a more accurate estimate of crash prediction, 10-fold cross-validation was adopted. This method is recognized for its susceptibility to yield minimal bias and variance in contrast with the other validation methods, including the leave-one-out method [43]. This type of strategies minimizes the impact of information dependence and improves the reliability of the resultant evaluation [44].

4 Results

The results of all SVM, MLP and BN models for each performance metric for all sets across the rain and snow weather conditions listed in Tables 1, 2. Each cell is filled with average of the relevant performance measure after 10-fold cross-validation. Based on the results, it can be seen in Table 1, MLP and SVM models depicted predictive performance during rainy conditions; SVM showed high accuracy with a value of 90.03% whereas MLP exhibited 93.44% and 86.21% in terms of recall and precision respectively. As for snow predictions (Table 2), MLP demonstrated the best accuracy and recall with values of 91.07% and 92.01% respectively, while SVM presented the best precision with 85.12% score. Even though BN depicted the lowest performance values with comparison to SVM and MLP, BN has demonstrated fairly good scores with performance in general over 80%. To further explore the findings visually, the aforementioned results are presented in Figs. 6 and 7. The overall performance values are presented in Fig. 8 which illustrates that superior predictive performance of MLP in terms of all adopted metrics followed by SVM then BN. This clearly indicates that, in this context, the use of MLP is preferable.

Table 1. Performance metrics for balanced crash events in rain weather.

Model	Set	Accuracy	Recall	Precision
SVM	Set 1	88.04	90.01	83.20
	Set 2	87.65	85.12	83.77

(continued)

Table 1. (*continued*)

Model	Set	Accuracy	Recall	Precision
	Set 3	**90.03**	91.21	84.80
MLP	Set 1	86.10	**93.44**	82.20
	Set 2	89.98	90.07	83.08
	Set 3	88.36	89.76	**86.21**
BN	Set 1	84.13	87.47	80.19
	Set 2	84.09	89.91	79.97
	Set 3	85.55	87.26	81.82

Table 2. Performance metrics for balanced crash events in snow weather.

Model	Set	Accuracy	Recall	Precision
SVM	Set 1	79.10	88.45	84.22
	Set 2	81.55	86.18	84.54
	Set 3	85.30	90.34	83.05
MLP	Set 1	88.92	90.49	**85.12**
	Set 2	**91.07**	**92.01**	84.86
	Set 3	90.37	90.77	83.33
BN	Set 1	88.21	86.13	79.98
	Set 2	88.29	85.39	80.07
	Set 3	87.19	87.65	82.11

Most of the highest recall values were obtained with MLP with performance over 89.76%. Conventionally, precision and recall metrics have an inherent tradeoff as one comes at the cost of the other; when applied with SMOTE resampling strategy, the values of both metrics have been found to achieve higher levels, 86.21% as the best score for precision and 93.44% for recall, which all were obtained during rainy conditions.

Within this context, Table 3 depicts the average performance results for participants sets. As can be seen, the overall accuracy for novice drivers is over 85%, while recall and precision achieved 89.33% and 82.49% respectively. On the other hand, the prediction results for the drivers with experience less than 10 years attained an average accuracy of almost 87% along with 88.11% recall and 82.72% precision. In contrast, set 3 which includes participants of more than 10 years driving experience have achieved the highest prediction performance with a higher precision and an accuracy and recall over 87% and 89% respectively. As can be noticed, when comparing the recall and precision scores, it is evident that the recall values attained by the three models are, in general, much superior than precision. The higher the recall, the more accurate the prediction result as it represents the correct crash prediction overall actual crash records. Put differently, it

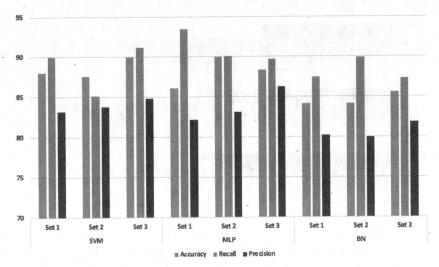

Fig. 6. Performance overview of proposed models in rain weather

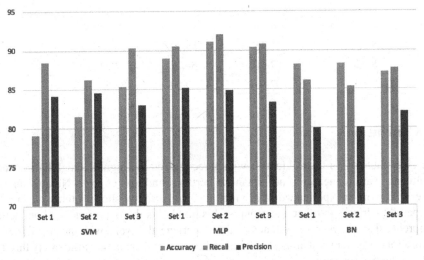

Fig. 7. Performance overview of proposed models in rain weather

depicts the proportion of crash occurrences that were correctly predicted by the models. The constructed models accomplished superb results for novice and experienced drivers.

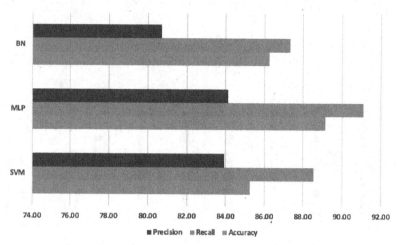

Fig. 8. Average performance results of proposed models

Table 3. Average results for novice drivers (Set 1) and experienced ones (Set 2 and 3).

Participants	Accuracy	Recall	Precision
Set 1	85.75	89.33	82.49
Set 2	87.11	88.11	82.72
Set 3	87.80	89.50	83.55

5   Conclusion

The assessment of traffic accidents is widely recognized as a significant global concern, resulting in various health issues, financial burdens, and loss of lives. Notably, novice drivers, who are inexperienced in driving, have consistently shown higher crash rates compared to drivers in other age groups. This highlights the importance of conducting comprehensive analyses for both novice and experienced drivers, with the goal of developing a robust system that aims to reduce road crashes and enhance traffic safety through highly effective strategies.

Researchers have provided evidence that crashes occurring in adverse weather conditions are more frequent compared to clear conditions. The impact of weather events on traffic maneuvers and safety outcomes has emerged as a significant concern, as drivers tend to adjust their driving behavior accordingly. However, there is a limited body of research that specifically focuses on evaluating crash events under rainy and snowy weather conditions for both experienced and inexperienced drivers. Furthermore, crash events are rare and unexpected, which can lead to misleadingly high prediction performance if the majority class dominates, while the crucial minority class is poorly represented. This study aims to address these issues by proposing an effective imbalance-learning approach for predicting road crash events among both novice and experienced drivers. The influence of rain and snow on traffic safety is investigated by analyzing real-time driver inputs, including throttle pedal position, brake pedal position, and wheel angle collected during various driving simulations. Additionally, an imbalance-learning strategy utilizing the SMOTE technique is implemented to develop robust machine learning models based on SVM, MLP and BN.

The findings of this study depict that machine learning models are capable of assessing weather-related crash events with high level of performance for novice and experienced drivers. Most of the highest performance values were obtained with MLP with recall performance over 93% and precision beyond 86%. Conventionally, all models have attained superior levels in terms of accuracy, recall and precision, however, MLP model appears to be the best performing classifier. This clearly indicates that, in this context, the use of MLP is preferable. Current ongoing work include additional data collection regarding inviting more participants for the experiment simulations and endorsing more profound comparisons in terms of types of road accidents while constructing more effective machine learning models based on fusion strategies. We aim to broaden the predictive performance of the machine learning models by including ensemble methods with several preprocessing strategies for feature selection and dataset sampling accordingly.

Acknowledgment. This research was jointly supported by (1) the Moroccan Ministry of Equipment, Transport and Logistics, and (2) the Moroccan National Center for Scientific and Technical Research (CNRST).

References

1. World Health Organization. WHO I Road Safety (2015). http://www.who.int/features/factfi les/roadsafety/en/
2. Zouhair, E.A.E., Mousannif, H., Al Moatassime, H.: Towards analyzing crash events for novice drivers under reduced-visibility settings: a simulator study. In: Proceedings of the 3rd International Conference on Networking, Information Systems & Security (2020)
3. Horswill, M.S., Hill, A., Buckley, L., Kieseker, G., Elrose, F.: Further down the road: the enduring effect of an online training course on novice drivers' hazard perception skill. Transp. Res. part F traffic Psychol. Behav. **94**, 398–412 (2023)
4. Ali, E.M., Ahmed, M.M., Wulff, S.S.: Detection of critical safety events on freeways in clear and rainy weather using SHRP2 naturalistic driving data: Parametric and non-parametric techniques. Saf. Sci., no. January 2018, (2019)
5. Yu, R., Zheng, Y., Abdel-Aty, M., Gao, L Exploring crash mechanisms with microscopic traffic flow variables: a hybrid approach with latent class logit and path analysis models. Accid. Anal. Prev., **125**, pp. 70–78 (2019)
6. Jahangiri, A., Berardi, V.J., MacHiani, S.G.: Application of real field connected vehicle data for aggressive driving identification on horizontal curves. IEEE Trans. Intell. Transp. Syst. **19**(7), 2316–2324 (2018)
7. Abou Elassad, Z.E., Mousannif, H., Al, H., Karkouch, A.: The application of machine learning techniques for driving behavior analysis : a conceptual framework and a systematic literature review. Eng. Appl. Artif. Intell. **87**, no. March 2019, p. 103312 (2020)
8. McDonald, A.D., Lee, J.D., Schwarz, C., Brown, T.L.: A contextual and temporal algorithm for driver drowsiness detection. Accid. Anal. Prev. **113**(January), 25–37 (2018)

9. Elamrani Abou Elassad, Z., Mousannif, H.: Understanding driving behavior: measurement, modeling and analysis. Adv. Intell. Syst. Comput. **5** (2019)
10. Pirdavani, A., et al.: Application of a rule-based approach in real-time crash risk prediction model development using loop detector data. Traffic Inj. Prev. **16**(8), 786–791 (2015)
11. Ba, Y., Zhang, W., Wang, Q., Zhou, R., Ren, C.: Crash prediction with behavioral and physiological features for advanced vehicle collision avoidance system. Transp. Res. Part C Emerg. Technol. **74**, 22–33 (2017)
12. Wang, C., Xu, C., Dai, Y.: A crash prediction method based on bivariate extreme value theory and video-based vehicle trajectory data. Accid. Anal. Prev., **123**, no. December 2018, 365–373 (2019)
13. Lee, D., Derrible, S., Pereira, F.C.: Comparison of four types of artificial neural network and a multinomial logit model for travel mode choice modeling. Transp. Res. Rec. J. Transp. Res. Board **2672**(49), 101–112 (2018)
14. Elamrani Abou Elassad, Z., Mousannif, H., Al Moatassime, H.: A real-time crash prediction fusion framework: an imbalance-aware strategy for collision avoidance systems. Transp. Res. Part C Emerg. Technol. **118**, no. November 2019, p. 102708 (2020)
15. Park, H., Haghani, A., Samuel, S., Knodler, M.A.: Real-time prediction and avoidance of secondary crashes under unexpected traffic congestion. Accid. Anal. Prev. **112**, no. May 2017, 39–49 (2018)
16. Parsa, A.B., Taghipour, H., Derrible, S., Mohammadian, A.: Real-time accident detection: coping with imbalanced data. Accid. Anal. Prev. **129**, no. May, 202–210 (2019)
17. Basu, S., et al.: Deep neural networks for texture classification—a theoretical analysis. Neural Netw. **97**, 173–182 (2018)
18. Li, Y., Chen, M., Lu, X., Zhao, W.: Research on optimized GA-SVM vehicle speed prediction model based on driver-vehicle-road-traffic system. Sci. CHINA Technol. Sci. **61**(5), 782–790 (2018)
19. Joo, Y.-J., Kho, S.-Y., Kim, D.-K., Park, H.-C.: A data-driven Bayesian network for probabilistic crash risk assessment of individual driver with traffic violation and crash records. Accid. Anal. Prev. **176**, 106790 (2022)
20. Elamrani Abou Elassad, Z., Mousannif, H., Al Moatassime, H.: Class-imbalanced crash prediction based on real-time traffic and weather data: a driving simulator study. Traffic Inj. Prev. 1–8 (2020)
21. Ameksa, M., Mousannif, H., Moatassime, H.A.L., Elassad, Z.E.A., Behavior, D., Driving, N.: Toward flexible data collection of driving behavior. XLIV, no. October, pp. 7–8 (2020)
22. Govinda, L., Raju, M.R.S.K., Shankar, K.V.R.R.: Pedestrian-vehicle interaction severity level assessment at uncontrolled intersections using machine learning algorithms. Saf. Sci. **153**, 105806 (2022)
23. Liu, J., Boyle, L.N., Banerjee, A.G.: Predicting interstate motor carrier crash rate level using classification models. Accid. Anal. Prev. **120**, no. November 2017, 211–218 (2018)
24. Zicat, E., Bennett, J.M., Chekaluk, E., Batchelor, J.: Cognitive function and young drivers : the relationship between driving, attitudes, personality and cognition. Transp. Res. Part F Psychol. Behav. **55**, 341–352 (2018)
25. Moral-García, S., Castellano, J.G., Mantas, C.J., Montella, A., Abellán, J.: Decision tree ensemble method for analyzing traffic accidents of novice drivers in urban areas. Entropy **21**(4), 360 (2019)
26. FHWA. How Do Weather Events Impact Roads? - FHWA Road Weather Management (2016). https://ops.fhwa.dot.gov/weather/q1_roadimpact.htm. Accessed 25 Jul 2019
27. Elamrani Abou Elassad, Z., Mousannif, H., Al Moatassime, H.: A proactive decision support system for predicting traffic crash events: a critical analysis of imbalanced class distribution. Knowledge-Based Syst. **205**, 106314 (2020)

28. Parsa, A.B., Movahedi, A., Taghipour, H., Derrible, S., Mohammadian, A.: Toward safer highways, application of XGBoost and SHAP for real-time accident detection and feature analysis. Accid. Anal. Prev. **136** (2020)
29. R. Akuh, M. Donani, S. Okyere, and E. K. Gyamfi, "The impact of perceived safety, weather condition and convenience on motorcycle helmet use: The mediating role of traffic law enforcement and road safety education," *IATSS Res.*, 2023
30. Malin, F., Norros, I., Innamaa, S.: Accident risk of road and weather conditions on different road types. Accid. Anal. Prev. **122**, no. August 2018, 181–188 (2019)
31. Slightly Mad Studios. Project CARS - THE ULTIMATE DRIVER JOURNEY. https://www.projectcarsgame.com/. Accessed 23 Jun 2019
32. Al, S., Dener, M.: STL-HDL: a new hybrid network intrusion detection system for imbalanced dataset on big data environment. Comput. Secur. **110**, 102435 (2021)
33. Chawla, N.V., Bowyer, K.W., Hall, L.O., Kegelmeyer, W.P.: SMOTE: synthetic minority over-sampling technique. J. Artif. Intell. Res. **16**, 321–357 (2002)
34. Kitali, A.E., Alluri, P., Sando, T., Wu, W.: Identification of secondary crash risk factors using penalized logistic regression model. Transp. Res. Rec. (2019)
35. Gosain, A., Sardana, S.: Farthest SMOTE: a modified SMOTE approach. In: Computational Intelligence in Data Mining, Springer, pp. 309–320 (2019). https://doi.org/10.1007/978-981-10-8055-5_28
36. Basheer, I., Hajmeer, M.: Artificial neural networks: fundamentals, computing, design, and application. J. Microbiol. Methods **43**(1), 3–31 (2000)
37. Schmidhuber, J.: Deep learning in neural networks: an overview. Neural Netw. **61**, 85–117 (2015)
38. Vapnik, V.N.: The Nature of Statistical Learning Theory. Springer New York (1995). https://doi.org/10.1007/978-1-4757-3264-1
39. Beryl Princess, P.J., Silas, S., Rajsingh, E.B.: Classification of road accidents using SVM and KNN. In: Advances in Artificial Intelligence and Data Engineering Springer, pp. 27–41 (2021). https://doi.org/10.1007/978-981-15-3514-7_3
40. Sirsat, M.S., Fermé, E., Câmara, J.: Machine learning for brain stroke: a review. J. Stroke Cerebrovasc. Dis. **29**(10), 105162 (2020)
41. Ramedani, Z., Omid, M., Keyhani, A., Shamshirband, S., Khoshnevisan, B.: Potential of radial basis function based support vector regression for global solar radiation prediction. Renew. Sustain. Energy Rev. **39**, 1005–1011 (2014)
42. Murphy, K.P.: Dynamic Bayesian Networks: Representation, Inference and Learning. University of California, Berkeley (2002)
43. Kohavi, R., Kohavi, R.: A Study of Cross-Validation and Bootstrap for Accuracy Estimation and Model Selection, pp. 1137--1143 (1995)
44. West, D.: Neural network credit scoring models. Comput. Oper. Res. **27**(11–12), 1131–1152 (2000)

Ontology-Based Mediation with Quality Criteria

Muhammad Fahad$^{(\boxtimes)}$ 🅙

Univ Lyon, Univ Lyon 2, UR ERIC, 5 Avenue Mendès France, 69676 Bron Cedex, France
f.muhammad@univ-lyon2.fr

Abstract. *Information integration* has a long history since humans started using and collecting information. But, it has been a strong focus of IT research since many recent years. It deals with providing a unified and transparent access to a collection of heterogeneous data sources. In information integration, the formulization of a global schema is a difficult task that manages multiple, autonomous and heterogeneous data sources. This paper presents a semantic system named *OntMed* for an ontology-based data integration of heterogeneous data sources to achieve interoperability between them. Our system is based on the quality criteria (*consistency, completeness and conciseness*) for building the reliable analysis contexts to provide an accurate unified view of data to the end user. The generation of an *error-free global analysis context* with the semantic validation of initial mappings generates accuracy, and provides the means to access and exchange information in semantically sound manner. In addition, data integration in this way becomes more practical for dynamic situations and helps decision makers to work within a more consistent and reliable *virtual data warehouse*.

Keywords: Information Integration · Ontology based Mediation · Ontology · Quality Criteria

1 Introduction

Information integration has a long history since humans started using and collecting information. But, it has been a strong focus of IT research since the 1970s. It deals with providing a unified and transparent access to a collection of heterogeneous data sources. In information integration, the formulization of a global schema is a difficult task that manages multiple, autonomous and heterogeneous data sources. According to the study by Bernstein and Haas [1], large enterprises utilize at least 40% of their budgets on information integration. A forecast about the market for the worldwide data integration and access software estimated to increase from $2.5 billion to $3.8 billion in 2007 to 2012 [2]. Although there is a huge amount of research done on this topic, but, it is still a hot issue in IT research. The two major research challenges are Intra-organization and Inter-organization information integration. *Intra-organizational data integration* is vital when different components of an organization adopt different systems to maintain. The need for Inter-organizational data integration is required in companies' merger, stock exchanges, etc. Research on this topic evolves with time from a centralized system with single and multiple data stores to managing federated data sources. Then, decentralized systems were designed where distributed integration is done by the application with or without central global schemas. Therefore, in the research literature, there are various approaches for information integration, such as data exchange, mediator-based, P2P data integration and exchange, data warehousing (DWH), etc., [3]. Data exchange aims at materialization of the global view by providing facilities of query answering without accessing the local data sources. There can also be P2P data integration and exchange between several peers. The approach is designed such that it allows queries over one peer, such that each of the peers is equipped with local and external sources.

When a data warehouse is overloaded with data, Chen and Zhao indentify some limitations such as flexibility, efficiency and scalability [25]. Therefore, an approach for data-information integration between *data cubes* is introduced to prevent hurdles in data maintenance and data analysis. There is also a need for data warehouse security techniques, such as data access control, which may not be easy to enforce and can be ineffective. Therefore, DWH also requires a cubic-wise balance method to provide privacy preserving range queries on data cubes in a data warehouse [26]. Along with privacy, quality of information is important as well. This paper is focused on *mediator based data integration* or *virtual warehouse based on quality criteria*. *Virtual* means that the data remains at the local data sources and a mediation layer provides *transparency* in answering querying by managing underlying local heterogeneous data sources.

The rest of the paper is organized as follows. Section 2 presents background terminologies for the reader. Section 3 discusses related work. Section 4 presents our *ontology-mediator based data integration* system *(OntMed)*. Section 5 presents the evaluation of our system with the help of Ontology Alignment for Query Answering (OA4QA) track of Ontology Alignment Evaluation Initiative (OAEI). Finally, Sect. 6 concludes our paper and shows future directions.

2   Background

This section presents the background concepts such as categories of DWH and mediation approaches. The first subsection presents three categories of DWH. The second subsection discusses different approaches for ontology-based data integration of heterogeneous data sources.

2.1   Data Warehouse and IT'S Categories

Inmon provides the first formal definition of a data warehouse as "A data warehouse is a subject-oriented, integrated, time-variant, and nonvolatile collection of data in support of management's decision-making process" [10]. In general, there are three categories of data warehouses [27].

Virtual View Approach. First, the virtual view approach deals when the physical data is stored in the local databases (dbs) and the repository of DWH only contains the schema of data. Query pre-processing and query shipping are required to answer queries which were made against the integrated view. This approach results in poor performance.

Materialized View Approach. Second, the materialized view approach deals when the data schema and the physical data both reside inside the repository of DWH. This approach leads to data management and maintenance problems due to large volumes of data.

Datamart Approach. Third, the datamart approach extracts data for a special purpose from the primary DWH for a datamart application. In this way, DWH only consists of limited knowledge and unable to manage huge amounts of data.

When data is stored in multi-dimensional form, it is visualized as a cube. When an enterprise implements datamart approach for building DWH, many cubes are formalized and each cube is an independent data aggregation. Each dimension in the cube depicts subsections of data which can be comparable and aggregatable. These independent cubes are isolated bits of information when investigated by analysts. They retrieve information from one single angle and not from a global view. This leads to problems like data duplication, inconsistency and query integrity [28]. For an example consider an example of data cubes illustrated in Fig. 1. The data of Cube Z consist of cube X and Cube Y, which may contain duplication of data and inconsistency issues may arise. Therefore, one must consider quality criteria when handling information integration of data cubes.

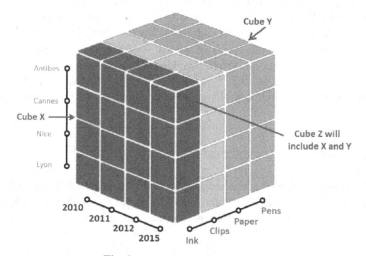

Fig. 1. An example of data cubes

2.2 Broad Categories of Ontology Based Information Integration

In research literature, there exist three approaches for querying heterogeneous data sources. According to the survey on ontology based information integration by Wache et al. [4], existing approaches can be classified into three broad categories. These different categories for ontology based data integration are represented in Fig. 2 and elaborated below.

Single Ontology Approach. In a single ontology approach, all source schemas are directly related to a shared global ontology. The shared global ontology provides a uniform interface and facilitates querying to the end-user. SIMS is a system that exploits a single ontology approach for querying heterogeneous ontologies [5].

Multiple Ontology Approach. In multiple ontology approach, each local data source is represented by its own local ontology. Local ontologies are mapped to each other on the basis of their similarities, i.e., their inter-ontology mappings are defined. The user query is rewritten for each local source and finally results are merged as per the inter-ontology mappings. OBSERVER is a system that exploits multiple ontology approach for querying heterogeneous ontologies [6].

Hybrid Ontology Approach. Hybrid ontology approach takes the advantages of both the above approaches [7]. For each local data source, there is a representative local ontology. All the local ontologies are mapped to a global shared ontology that provides a unified view. The advantage of this approach is that new data sources can be easily integrated without changing in existing mappings.

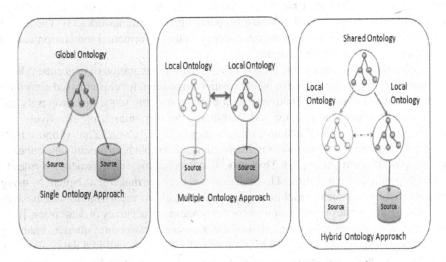

Fig. 2. Different approaches for Ontology based Data Integration

On the basis of these approaches, mapping between the local and global ontologies are specified. When the source schemas are defined with respect to the global schema, the approach is known as Local-As-View (LAV). But when the global schema is defined with respect to the local schema, the approach is known as Global-As-View (GAV). A mixed approach can also be defined as GLAV.

3   Related Work

Data integration is essential since the analysis context is built using data from different heterogeneous data sources and providing an end-user with the unified view of underlying data [8]. For the integration of heterogeneous data and on-line analytical processing, data warehousing has long been adopted that aims at building a centralized database which contains all (or selected) data originating from various heterogeneous data sources designed in the multidimensional manner [9]. Two well-known techniques for building warehouses are provided by *Inmon* [10] and *Kimball* [11]. In addition, on-line analytical processing (OLAP) systems may stock terabytes of data and handle queries from millions of end-users. Since in the data warehouse, heterogeneous data is warehoused that facilitates the decision processes, this approach is recognized by its performance due to its query response time. With the time when there is a change in data, updates are required by the decisional tools for the accurate decision making. The update process needs much additional cost for the refreshment of data. Data warehouse by mediation is thus increasingly seen as a solution to this problem that aims at building a virtual data warehouse to build analysis context on-the-fly.

Kharlamov et al. develop Ontology Based Data Access (OBDA) system which contributes a conceptual view of data and the ontology performs as a mediator between the user and the data [32]. It enables querying a database which is represented with ontology by abstracting away from the technical schema-level details of the local data sources. Later on, Can and Unalir reuse the Ontology Based Access Control (OBAC) model within the scope of OBDA which supports a privacy framework [33]. The aim of their work is ontology-based secure data access so that information integration can be performed in a privacy-aware manner.

Besides these, there are some works for information integration of data cubes. Kaur and Kaur implement cost-effective parallelization techniques for cluster based algorithm to compute data cubes [24]. Their approach exploits existing sequential proposals and is intended to target load balance and formulates the communication effectively and efficiently. Similarly to facilitate information integration, Chen and Zhao propose rule-based cube exploration by introducing inter-relationships that hold among data cubes to avoid loss of data semantics [25]. They mix OLAP technology with association rules to tackle the problem of navigation of large data so that decision makers can benefit by using low capacity devices such as mobile devices. Liu et al. also present a cubic-wise balance approach for the privacy preservation and better range query accuracy in data cubes. Their approach provides a precise estimation of the answer for range-sum queries, enabling preservation of the confidential information of individual data cell in a data cube [26]. Huang et al. propose a semantic cube model approach for data warehouse enhancement to reduce the duplication of data and to improve the performance of query integrity [28]. An end-user designs the generalization relationship between different cubes. To reap the targeted benefits, they exploit connectivity by identifying the relationship between data cubes.

Ontology-based information integration is being used in multiple domains. Sobral et al. propose an ontology-based framework to support integration and visualization of data from intelligent transportation systems [29]. They develop Visualization-oriented Urban Mobility Ontology as a semantic pillar for knowledge-assisted visualization tools. Their domain ontology consists of three facets which interrelate the characteristics of spatio-temporal mobility data, visualization mechanisms and expert knowledge. Belitz-Hellwich implements information exchange between a logistics simulation model and a finite element model through ontologies [30]. They connect source models of both domains, build interfaces between the models and the ontologies, and design an application that executes the exchange of information between them.

Now-a-days, the use of ontologies for describing semantics of data in data warehouse design has revolutionized the performance in terms of accuracy and provides means to exchange information in semantically sound manners in heterogeneous environments. Although there is already a lot of research in this area, there are still many issues that need to be resolved, especially concerning the quality of results. Based on the previous research [12], we analyzed various situations when local ontologies are merged together that need inconsistency, incompleteness and inconciseness checking. Therefore, we build a quality criteria based on *consistency, completeness and conciseness* parameters for building the reliable analysis context in the data warehouse design based on a mediation approach. We further investigated our approach that exploits the semantic validation of initial mappings from the information present in the local ontologies that form global analysis context on-the-fly. Global ontology as an analysis context provides a conceptual unified view to the end-user about the underlying data. The user performs a query over the data. As this query is expressed in a global schema terminology, it must be reformulated in terms of local data sources such that it can be executed. Otherwise, query results are incomplete or null. Once the results are retrieved from the global schema, they need to be merged together and presented to the user in terms of a global terminology. In this whole process, we believe that the generation of error free global analysis contexts with semantic validation of initial mappings would generate more accuracy, provide means to exchange information in semantically sound manners, data integration by this way becomes more practical for dynamic situations and helps decision maker to work within more consistent and reliable virtual data warehouse. There are many other problems under this topic, such as data extraction, cleaning, reconciliation, and optimization of query answers, but these are out of scope of this paper. Our work is limited to identification of mappings between local data sources, validation of mappings based on quality criteria, generation of global schema, answering queries formulated with the vocabulary of the global schema and aggregation of final results.

4   Ontology-Based Data Integration

Traditional data warehouse systems based on *ETL* mechanisms *extract, transform,* and *load* data from several heterogeneous data sources into a single queryable data. This approach of data integration is tightly coupled because the data is located in the same repository during the execution of query, therefore provides fast query response. As of modern research (almost from 2009 onward), the trend in data integration has favored loosening the coupling between the data. This requires mediation based uniform query-interface, where an end-user can write a query, and the mediation layer transforms it into specialized queries over the original databases, and retrieves the results. One example of a data warehouse based on mediation approach is presented by [13].

A data warehouse based on a mediator approach consists of three main elements; *data source schemas* as local ontologies, *a global ontology* and *correspondence rules* between local-global ontologies for the query answering. Building a global ontology is essential as it allows the execution of decisional queries on underlying sources, query conversion from the global schema vocabulary to the data sources vocabulary, and building the data cube on-the-fly from the obtained results from the different data sources. But, the mediation layer has to face a number of problems; since local ontologies may represent the same knowledge in different ways producing various mismatches while the construction of a global schema. In addition, it is very hard to perform their manual integration beyond a certain degree of size (number of concepts, properties and relations), complexity of axioms, and number of ontologies. Therefore, it needs fully automatic techniques for supporting interoperability in dynamic environments such as building virtual data warehouse where analysis context is made on-the-fly. This dilemma requires the need of a reliable ontology merging approach that should be capable enough to find semantic heterogeneity in source ontologies and resolve it automatically to produce accurate, consistent and coherent global merged ontology on-the-fly based on a number of local source ontologies. Since it is in the analysis and decision domain, validation of initial mappings that form global schema is more than a challenging task for decision making and sound manipulation of data. Therefore, we took this initiative to integrate quality criteria inside the mediation layer to build a reliable analysis context to achieve correctness of desired query results.

4.1   Representative Schema for Local Data Sources

According to Haw et al. (2017) the mapping between a database and an ontology is regarded as a case of data integration [31]. Concepts of the ontology are mapped on the relevant entities in the db. We believe each of the data sources has some representative schema, i.e., local source ontologies already exist. If not, use some tool to generate individual ontologies [14, 15].

4.2   Generation of Global Schema

The second step of ontology-based data integration is to formulate the global schema. Our approach is flavored by building a mediator based data integration that facilitates to physically reside data in original sources. It minimizes the burdens for end-users from locating sources significant to a query, interacting with each one in isolation, and manually integrating data from multiple data sources. For our ontology-based data integration system, we use our previously designed ontology merging system DKP-AOM that deals with the problem of providing a unified view (in terms of global ontology) of local ontologies which represent distributed and heterogeneous data sources. Further information about our DKP-AOM system can be found in [16]. The user creates a query by using vocabulary of the global schema and the mediator exploits some techniques to carry out the query to achieve meaningful answers. For this, it needs to translate a query expressed with the mediated schema terminology into one that refers directly to the local schema.

4.3   A Novel Framework for Data Integration with Reliable Mediation

We design a novel framework for the data integration with reliable mediation named *OntMed* based on a hybrid approach; where local ontologies represent local data sources, and our DKP ontology merging framework generates a global ontology automatically from the local ontologies. Finally a query engine executes query and retrieves result by query rewriting, obtains local answers, and constructs global result. Figure 3 depicts the data integration approach adapted by our system *OntMed*.

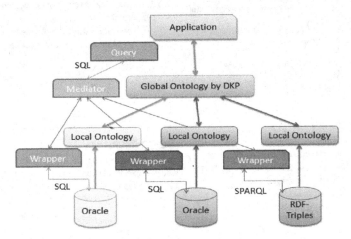

Fig. 3. Data Integration approach adapted by OntMed

It is well argued in the research area that often the generated results of data integration suffer from inconsistency, incompleteness and redundancy. Therefore, we implemented the quality criteria based on the Consistency, Completeness and Redundancy parameters [17, 18] in a mediation layer to build an accurate analysis context so that execution of simple queries on the global ontology reveals correct results. Figure 4 depicts our quality criteria for achieving reliable mediation and also summaries various errors. The important thing which should be considered is that global ontology should be free from these errors as it serves an analysis context on which user queries are performed and final result is aggregated. The details about these criteria are as follows.

Inconsistency in the Global Ontology. Inconsistency in the global ontology means that there exists some disagreement within the ontology. It also means there exists some conflicts or sort of contradictory knowledge inferred from the concepts, definitions and instances within the ontology. Inconsistency in the global ontology produces ambiguities, contrary axioms, contradictions in final results and compromises precision. There are mainly three types of errors that can cause inconsistency and ambiguity in the ontology. These are Circulatory errors, Partition errors and Semantic inconsistency errors.

Incompleteness in the Global Ontology. Incompleteness occurs when the global ontology contains various types of domain knowledge in the form of concepts, properties and definitions, but overlooks some of the important information about the domain. The incompleteness of the domain knowledge often creates ambiguity, and lacks reasoning and inference mechanisms. The incompleteness errors are due to incomplete concept

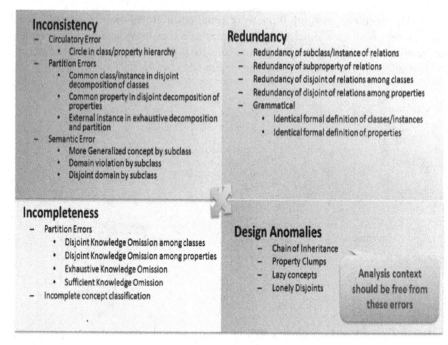

Fig. 4. Quality criteria for Reliable Mediation

specification and partitions errors due to disjointness and exhaustive knowledge omission between concepts.

Redundancy in the global ontology. Another important aspect is to make global ontology concise, so that it stores only necessary and sufficient knowledge about the concepts. Inconciseness or repetition or redundancies in global ontology not only compromise usability and convenience, but also create problems for the maintenance, preservation and manageability of concepts and properties within ontologies. Generally, there are redundancy errors when particular information is inferred more than once from the relations, classes and instances found in the global ontology.

Design Anomalies in the Global Ontology. Besides taxonomical errors, Baumeister and Seipel pointed out some anomalies based on bad designs. These anomalies restrict simplicity, coherence and maintainability of taxonomic structures within the ontology [19]. These anomalies do not generate incorrect reasoning from the axioms, but highlight controversial, problematic and poorly designed areas in ontology that cause issues of maintainability. Therefore it is advisable to detect and repair or remove these anomalies. Ontologies free from these anomalies improve the usability, clarity and provide better maintainability over the semantic web.

Once the global ontology is generated by the DKP ontology merging system, this serves as an analysis context for our data integration system. We have participated in the conference and OA4QA track of OAEI 2015 and results of our system are competitive, see [20]. Figure 5 illustrates the execution of different steps of our system for the data

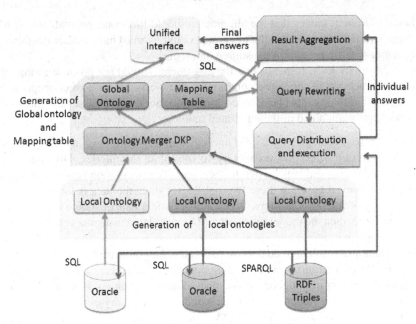

Fig. 5. Execution of different steps by OntMed

integration. We have tested our approach of data integration on a case study related to Shopping Mall data sources which are modeled in two instances of Oracle. These data sources are heterogeneous and separately developed. In addition, we blend our case study with the RDF triple data source, to add more heterogeneous environment in the case study. However, we have incorporated very limited information about the RDF data source in this case study, and mainly it circulates over Oracle data sources. In the near future, we will present this case study.

5 Validation of OntMed Methodology

We implement OntMed (ontology-based data integration via mediation) module as a part of an Ontology Merging System (DKP-AOM). For the evaluation of our system, we choose the Ontology-Based Data Access (OBDA) scenario (see Fig. 6). This scenario is considered as a part of Ontology Alignment for Query Answering (OA4QA) track of Ontology Alignment Evaluation Initiative (OAEI) 2015 [21]. Although it is old, it serves its purpose to validate the methodology. There are three rules which should be taken into account to reduce the number of potentially unintended consequences [22]. The following are these rules:

(i) *Consistency Rule.* It means that the alignments should not bring unsatisfiable classes in the integrated merged ontology. Only consistent alignments should be detected between source ontologies which lead to consistent integrated ontology.

(ii) *Locality Rule.* It means that the alignments should link concepts that have similar neighborhoods. Concepts in the integrated ontology must have similar neighboring concepts that are present in the source ontologies.

(iii) *Conservativity Rule.* It means that the alignments should not propose changing in the classification of the source ontologies. New classification of concepts in the integrated ontology must be avoided and preservation of the same classification of concepts is maintained in the integrated ontology.

There are two ontologies (i.e., QF-Ontology and DB-Ontology) in the OBDA scenario. QF-Ontology offers the vocabulary to formulate the queries. DB-Ontology is connected to the data and it is hidden to the users. The integration based on proposed mappings is achieved between these two ontologies since the data is linked only with the vocabulary of the DB-Ontology. When a query and a pair of source ontologies are provided as input, then the model result set will be calculated using the correspondences for the ontology pair.

The accuracy of resultant model answer sets are measured based on the Precision and Recall [23]. The evaluation of ontology alignment based on a set of reference alignments are taken into account investigating this aspect of data integration. Precision and recall values are computed by initiative campaign based on the capacity of the generated mappings to answer a set of queries in an OBDA scenario in presence of multiple ontologies. The track is intended to study the practical consequences of logical violations disturbing the detected mappings. In addition, they compare the different repair techniques adopted by the participant ontology matching systems.

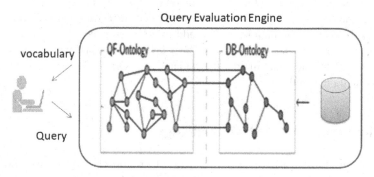

Fig. 6. Ontology-based data access scenario

The well-known conference data set is used to promote understanding of the dataset and the queries. However, it is extended with synthetic ABoxes. The results of participated systems are taken into account by the SEALS platform. The evaluation results are generated on available reference alignments of Conference track (RA1) and its repaired version (RAR1). Table 1 illustrates the results for the whole set of queries showing the average precision, recall and f-measure. All the participant Ontology matchers are tested on 18 queries, for which the sum of expected answers is 1724. There is only one answer for some queries. But some queries have as many as 196 answers. The details of participant ontology matching systems, evaluation criteria and evaluation results can be found in their individual research publications. However, the following points highlight the evaluation results showing the salient features of our tool.

- The alignments generated only by the systems (AML, DKPAOM, LogMap, LogMap-C and XMap) allowed answering *all the queries* of the evaluation.
- According to jury of campaign, the best global results have been achieved for violations queries. Ontology Matching tools (AML, DKPAOM, Log- Map, COMMAND, LogMapC and XMAP) generate excellent correct results, in decreasing order of f-measure w.r.t. RA1.
- Our system DKPAOM obtained a remarkable f-measure of 0.999 w.r.t. RAR1 (see evaluation results aggregated by query family[1]). It means logical violations are successfully handled by our tool.

Table 1. Global Evaluation Results published by OA4QA track **taken from the original paper [21].

Matcher	Answered Queries	ra1			rar1		
		Prec	Rec	F-m	Prec	Rec	F-m
AML	18/18	0.778	0.750	0.759	0.757	0.750	0.746
COMMAND	14/18	0.557	0.611	0.575	0.536	0.611	0.562
CroMatcher	12/18	0.464	0.479	0.434	0.407	0.458	0.404
DKPAOM	18/18	0.667	0.618	0.635	0.666	0.639	0.648
GMap	9/18	0.324	0.389	0.343	0.303	0.389	0.330
JarvisOM	17/18	0.409	0.430	0.386	0.407	0.430	0.385
Lily	11/18	0.389	0.368	0.376	0.389	0.389	0.389
LogMap	18/18	0.750	0.750	0.741	0.729	0.750	0.728
LogMapC	18/18	0.722	0.694	0.704	0.722	0.694	0.703
LogMapLite	11/18	0.409	0.423	0.379	0.351	0.402	0.348
Mamba	14/18	0.556	0.528	0.537	0.556	0.528	0.537
RSDLWB	6/18	0.131	0.180	0.127	0.129	0.180	0.126
ServOMBI	6/18	0.222	0.222	0.222	0.222	0.222	0.222
XMAP	18/18	0.778	0.675	0.702	0.720	0.654	0.671

6 Conclusion

Data integration is crucial since the analysis context is built using data from different heterogeneous sources. With the grown usage of data warehouse, the question about the usage of a mediation approach for building analysis context on-the-fly has become

---

[1] https://oaei.ontologymatching.org/2015/oa4qa/results.html.

even more important in today's dynamic world. Although, there has been already a lot of research in this area, there are still many issues that need to be resolved especially concerning with the quality of results. There are various points about the ontology merging and conceptual schema merging. In general, an ontology is a broad concept than conceptual schema and a conceptual schema is regarded a sub-concept of ontology. Therefore, merging conceptual schema requires less effort than ontology merging. Integration of databases has been the focus of years of research. There are many approaches in the research literature for the data integration with/without the support of ontologies. Ontologies conceptualize concepts with their generalization and formulate properties and axioms to equip the semantics. The relational model on the other hand provides limited semantic description about the data. For example, there is no or very limited generalization support, no axiomatic definitions, etc. An absence of the rich semantics also poses a difficulty in recognition of concepts during their mappings and hence their merging raises the level of difficulty.

This paper contributes an ontology-based data integration framework. Integration of data sources also depends on the representation and quality of its representative schema or ontology. The more representation model is good, the more integration is easily done with quality. The presented approach based on the quality criteria (consistency, completeness and conciseness) for building reliable analysis contexts in data warehouse design suits well and provides accurate unified view of data to the end user. In addition, our approach builds the virtual data integration environment with the least human intervention. Our automatic ontology merger generates the global ontology from the local ontologies. Our approach exploits semantic validation of initial mappings from the knowledge that is present inside the local ontologies that form global analysis context on-the-fly. The generation of an error-free global analysis context with the semantic validation of initial mappings generates more accuracy, and provides means to exchange information in semantically sound manners. In addition, data integration by this way becomes more practical for dynamic situations and helps decision maker to work within more consistent and reliable virtual data warehouse.

One of the future directions is to present a case-study on ontology based information integration where we demonstration how various errors in the mapping phase create hurdles achieving correctness and accuracy in the integration process and aggregation of final results. Other is to evaluate our framework on a recent dataset and report its performance.

Acknowledgement. The research depicted in this paper is funded by the French National Research Agency (ANR), project ANR-19-CE23–0005 BI4people (Business Intelligence for the people).

References

1. Bernstein, P.A., Haas, L.M.: Information integration in the enterprise. Commun. ACM (CACM) **51**(1), 72–79 (2008)
2. Ziegler, P., Dittrich, K.R.: Three Decades of Data Integration - All Problems Solved?, WCC, pp. 3–12 (2004)

3. IDC, Worldwide Data Integration and Access Software, 2008–2012, Forecast. Doc No. 211636, Apr. (2008)
4. Wache, H., et al.: Ontology-based integration of information - a survey of existing approaches. In: Proceedings of the IJCAI-01 Workshop on Ontologies and Information Sharing (2001)
5. Arens, Y., Knoblock, C.A., Hsu, C.: Query Processing in the SIMS Information Mediator. In The AAAI Press (1996)
6. Mena, E., Kashyap, V., Sheth, A.P., Illarramendi, A.: OBSERVER: an approach for query processing in global information systems based on interoperation across pre-existing ontologies. In: Proceedings of the 1st IFCIS International Conference on Cooperative Information Systems (CoopIS 1996), pp. 14–25 (1996)
7. Cruz, I.F., Xiao, H.: Using a layered approach for interoperability on the semantic web. In: Proceedings of the 4th International Conference on Web Information Systems Engineering (WISE), pp. 221–232, Rome, Italy (2003)
8. Lenzerini, M.: Data Integration: A Theoretical Perspective. PODS, pp. 233–246 (2002)
9. Chaudhuri, S., Dayal, U.: An overview of data warehousing and OLAP technology, ACM SIGMOD Record **26**(1), 65–74 (1997)
10. Inmon, W.H.: Building the Data Warehouse. John Wiley & Sons Inc, New York, USA (1992)
11. Kimball, R.: The operational data warehouse. DBMS **11**(1), 14–16 (1998)
12. Fahad, M., Qadir, M.A.: A framework for ontology evaluation. 16th ICCS Supplement Proceeding, **354**, pp.149–158, France (2008)
13. Maiz, N., Fahad, M., Boussaid, O., Tayab, F.B.: Automatic ontology merging by hierarchical clustering and inference mechanisms. In: proceedings of 10th International Conference on knowledge Management and Knowledge Technologies (I-Know'10), Sept 1–3, Messe Congress Graz, Austria (2010)
14. Upadhyaya, S.R., Kumar, P.S.: ERONTO: a tool for extracting ontologies from extended E/R diagrams. ACM Symposium on Applied Computing (2005)
15. Xu, Z., Zhang, S., Dong, Y.: Mapping between relational database schema and OWL ontology for Deep Annotation. In: Proceedings of the 2006 IEEE/WIC/ACM International Conference on Web Intelligence (WI'06), IEEE (2006)
16. Fahad, M., Moalla, N., Bouras, A.: Detection and resolution of semantic inconsistency and redundancy in an automatic ontology merging system. J. Intell. Inf. Syst. (JIIS) **39**(2), 535–557 (2012)
17. Gómez-Pérez, A.: Evaluating ontologies: cases of study. IEEE Intell. Syst. and their Appl. **16**(3), 391–409 (2001)
18. Gomez-Perez, A., Fernández-López, M., Corcho, O.: Ontological engineering: with examples from the areas of knowledge management. E-Commerce and the Semantic Web. Springer, London (2004). https://doi.org/10.1007/b97353
19. Baumeister, J., Seipel, D.S.: Smelly owls–design anomalies in ontologies. In: 18th Intl. Florida AI Research Society Conference, pp. 251–220. AAAI Press (2005)
20. Fahad, M.: Initial results for ontology matching workshop 2015, DKP-AOM: Results for OAEI 2015. In: CEUR Workshop Proceedings 1766, pp. 82–96. 5 (2015). http://oaei.ontolo gymatching.org/2015/conference/index.html
21. Cheatham, M., Dragisic, Z., Euzenat, J., Faria, D., Ferrara, A., et al.: Results of the ontology alignment evaluation initiative 2015. 10th ISWC Workshop on Ontology Matching (OM), Oct, Bethlehem, United States. pp. 60–115 (2015)
22. Solimando, A., Jiménez-Ruiz, E., Guerrini, G.: Detecting and correcting conservativity principle violations in ontology-to-ontology mappings. In: Mika, P., et al. (eds.) ISWC 2014. LNCS, vol. 8797, pp. 1–16. Springer, Cham (2014). https://doi.org/10.1007/978-3-319-119 15-1_1
23. Solimando, A., Jimenez-Ruiz, E., Pinkel, C.: Evaluating ontology alignment systems in query answering tasks. Poster paper at International Semantic Web Conference (ISWC) (2014)

24. Kaur, P., Kaur, P.: New approach of computing data cubes in data warehousing. Int. J. Inf. Comp. Technol. **4**(14), 1411–1417 (2014)
25. Chen, Z., Zhao, T.: A cube model approach for data warehouse. Proceedings of the 4th International Conference on Mechatronics, Materials, Chemistry and Computer Engineering, pp. 846–849 (2015)
26. Liu, Y., Sung, S.Y., Xiong, H.: A cubic-wise balance approach for privacy preservation in data cubes. Inf. Sci. **176**, 1215–1240 (2006)
27. Alejandro, G.T., Marotta, A.: An Overview of Data Warehouse Design Approaches and Techniques (2001)
28. Huang, S., Chou, T., Seng, J.: Data warehouse enhancement: a semantic cube model approach. Inf. Sci. **177**(11), 2238–2254 (2007)
29. Sobral, T., Galvão, T., Borges, J.: An ontology-based approach to knowledge-assisted integration and visualization of urban mobility data. Expert Systems with Applications, **150** (2020)
30. Belitz-Hellwich, W.: An Ontology-Based Platform for Information Integration; Supporting Sustainable Smart Transportation Infrastructure (2023). https://www.diva-portal.org/smash/get/diva2:1737543/FULLTEXT01.pdf
31. Haw, S.-C., May, J.-W., Subramaniam, S.: Mapping relational databases to ontology representation: a review. In: ICDTE'17, pp. 54–55 (2017)
32. Kharlamov, E., Hovland, D., Jimenez-Rui, E., et al.: Ontology based data access in statoil. J. Web Semantics **44**, 3–36 (2017)
33. Can, O., Unalir, M.: Revisiting ontology based access control: the case for ontology based data access. In: Proceedings of the 8th International Conference on Information Systems Security and Privacy (ICISSP), pp. 515–518 (2022)

Predicting Driving License Applicant's Performance for Car Reverse Test System

Abderrahman Azi[✉] [iD], Abderrahim Salhi [iD], and Mostafa Jourhmane [iD]

Information Processing and Decision Laboratory, Sultan Moulay Slimane University, Beni-Mellal, Morocco
abderrahman.azi@gmail.com, m.jourhmane@usms.ma

Abstract. This research is focuses on evaluating the performance of applicants during driving license exam. The study utilizes the generated paths from the last developed part of the Car Reverse Test (CRT) system, which are categorized accordingly. The main objective is to interpret the path chosen by applicants and their vehicle control techniques in order to predict a score to be proposed to the supervisor. The collected applicant data are classified using data mining techniques using Multi-Layer Perceptron (MLP), Naïve Bayes (NB), Support Vector Machine (SVM) and Decision Trees algorithms. The classification results are generally acceptable, with the MLP classifier achieving the highest accuracy percentage of 93% compared to other classifiers. Thus, this technique is strongly recommended regarding to its accuracy, followed by NB, SVM and DT, which achieved classification accuracies of 88%, 80% and 67% respectively.

Keywords: Driving license exam · CRT system · Multi-Layer Perceptron · test score prediction · Naïve Bayes · Support Vector Machine · Decision Trees

1 Introduction

In Morocco, improving several services related to transportation, promoting equality among citizens, and ensuring administrative transparency in the departments responsible for issuing driving licenses are among the priorities of the government authorities. Undoubtedly, obtaining a driving license is an extremely important task due to its close association with road safety. Providing facilities in this area, regardless of their type, can hamper efforts to combat the scourge of road accidents and help individuals who lack the necessary qualifications to obtain a driving license.

Issuing a driving license is not just an ordinary administrative procedure. Moreover, it represents a mechanism that leads to objectively evaluating the theoretical and practical training received by the candidate in order to serve a fair ability judgment and maximum transparency. The strictness and seriousness that must surround the process of obtaining a driving license require supervisors to demonstrate competence, integrity, and accurate application of the legal requirements applicable in this field.

The implementation of these requirements must pass through two levels, as evidenced by studies conducted by the Moroccan government in this field: The first level involves

R. El Ayachi et al. (Eds.): CBI 2023, LNBIP 484, pp. 89–102, 2023.
https://doi.org/10.1007/978-3-031-37872-0_7

ensuring that candidates meet the legal and regulatory requirements necessary to pass the driving license exam across different categories. The second level pertains to the methods employed for conducting the theoretical and practical tests, which can be summarized as follows:

For the theoretical test, it is imperative to ensure the supervision of test sessions with utmost integrity, while avoiding any behaviors that could potentially impact the accurate evaluation of results. Conversely, the practical test should be conducted with transparency, fairness, and equality. Thus, an assessment based on objective criteria is crucial, enabling supervisors to evaluate candidates possessing the required qualifications and competence to execute various maneuvers within and outside urban areas, as elucidated by the authors in [1].

Since 2005, the Moroccan government has implemented a reform strategy for the driving license exam, aiming to incorporate technology to enhance the efficiency of each component of the exam. Presently, the theoretical test is conducted using a machine, while the practical test has remained unchanged. In light of this, we propose the implementation of the 'Driving License Supervisor System' (DLES) project. The primary objective of this system is to address the challenge of automatically supervising the practical portion of the driving license exam. The DLES project intends to utilize a video surveillance architecture, as described in [1], along with various image processing techniques to achieve improved outcomes.

Up to this stage, we have focused on the car reverse test, for which a sub-system named Car Reverse Test system (CRT system) is proposed. For development purposes, we have divided the proposed sub-system into three parts:

In the first part [1], the objective was to propose a motion detection method using a linear approach named the line area monitoring algorithm (LMA). This method would allow the program to determine when a vehicle has crossed the start or end line, marking the beginning or end of the supervising process. The advantage of using the proposed method is to minimize processing of static scenes, accelerate video event analysis, and improve storage efficiency.

In the second part [2], we have proposed a method based on object detection and tracking techniques to define the borders of the vehicle and then detect its center point while it trespasses the tracking area. At the end of the process, when the vehicle crosses the end line, the program generates the path to be stored for further treatment. Moreover, the system successfully detects when the vehicle crosses the left and right lines, indicating that the candidate has failed the test.

In this paper, another part of the CRT system is studied. The system's focus is on processing the trajectory followed by the vehicle in the tracking area and objectively evaluating the candidate's behaviour during the test. The main objective is to predict a score for this test using a machine learning technique and then propose it to the supervisor. The system should allow the supervisor to validate the score. If the score offered by the supervisor differs from that proposed by the system, it will be considered in the learning process.

We have organized the rest of this paper as follows: In Sect. 2, we investigate some related works. Then, in Sect. 3, we describe the methodology. In Sect. 4, we report and discuss results and program performance. Finally, Sect. 5 concludes the paper and outlines the future work.

2 Related Works

In this section, we want to discuss related works, delve into the relevant literature, and highlight the variations, as well as propose potential avenues for future research. Among many issues, proposing systems that make sense of vehicle trajectories is a subject of many studies, such as [3–5]. In general, these studies use data obtained from video surveillance image segmentation or GPS data. Sequences of special positions detected over time identify trajectory data. Therefore, a raw trajectory consists of a sequence of n points: $T = \{p_1, p_2, ..., p_n\}$, in which $p = \{x, y, t\}$, where x and y represent the position of a moving object in space at a specific moment in time t.

Many methods for trajectory classification exist in the literature. However, they do not propose any new types of classifiers. The main idea is to focus on extracting the most accurate features for classification. These features define any type of data extracted from the entire trajectory or a slice of it using segmentation methods, statistics, or any other technique. Then, it is time to use a classifier such as neural networks (NN), random forests (RF), support vector machine (SVM), multilayer perceptron (MLP), etc.

The authors of [6] introduced one of the first techniques for trajectory classification, in which an SVM classifier is used to evaluate results obtained from many types of datasets (Animals, Vessels, Hurricanes, and Synthetic Dataset).

Many studies have focused on the classification of trajectory data of transportation modes. For instance, authors of [7] and [8] used SVM as a classification technique, while the authors of [9] introduced Convolutional Neural Networks (CNNs) for the same purpose. Similarly, the authors of [10] introduced Random Forest (RF) and Decision Tree (DT), while the authors of [11] and [12] employed MLP and K-Nearest Neighbors (KNN). Lastly, the authors of [13] employed Naïve Bayes and Quadratic Discriminant Analysis as classification techniques.

Table 1 summarizes the related works for trajectory classification, including datasets used by each method with their associated accuracies.

Table 1. Summary of previous research

Technique	Classifier	Datasets	Accuracy (%)
Lee (2008) [6]	SVM	Animals, Vessels, Hurricanes and Synthetic Dataset	75.73

(continued)

Table 1. (*continued*)

Technique	Classifier	Datasets	Accuracy (%)
Lee (2011) [8]	SVM	Taxis from San Francisco, Synthetic Dataset	89.84
Bolbol (2012) [7]	SVM	Private Dataset	88
Tragoupoulou (2014) [10]	RF, DT	Private Dataset	92.81, 87.61
Varlamis (2015) [11]	RF, DT, KNN, SVM, MLP	Private Dataset	65.80, 67.13, 84, 75.5, 87.2
Xiao (2017) [12]	KNN, DT, SVM, RF	Geolife	74.44, 86.05, 87.69, 90.29
Dabiri (2018) [9]	CNN	Geolife	84.8
Etemad (2018) [13]	RF, DT, Bayes, MLP, Quadratic Discriminant Analysis	Geolife	85.56, 88.07, 85.18, 63.30, 54.76

3 Methodology

In this section, we will introduce the methodology adopted to predict a candidate's score for the car reverse test by analyzing the generated car trajectory, as discussed in our previous paper [2]. Figure 1 shows an example of a generated path from an offline video sequence.

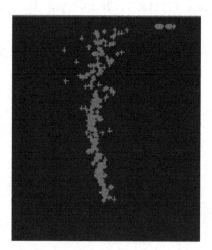

Fig. 1. The car trajectory generated by CRT system [2]

Following the approach of the authors of [14] and discussed in [15] and [16], as shown in Fig. 2, involves four distinct phases: data collection, data processing, data mining and data interpretation.

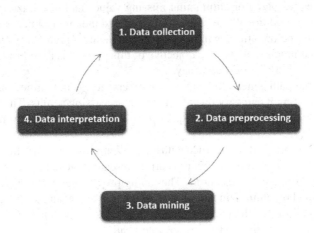

Fig. 2. System process model applied to CRT system adapted from authors of [14]

3.1 Data Collection

This research uses a dataset composed by trajectories made by driver's license applicants during car reverse test. The number of applicants for this study was N = 50, representing the number of candidates who passed the exam. In this stage, we have applied the process to offline videos of real scenes.

For testing purposes, the supervisor proposes a score for each test, represented by a number between 5 and 10. Therefore, this score will be associated with a class. In this study, we chose to categorize applicants into five classes with labels ranging between 0 and 5.

To provide a clearer view of the dataset, Table 2 represents the class labels along with the number of applicants and the associated percentage for each class.

Table 2. Class labels according to applicants scores

Score	Class	Number of candidates	Percentage
5	0	8	16%
6	1	10	20%
7	2	11	22%
8	3	10	20%
9	4	6	12%
10	5	5	10%

3.2 Preprocessing and Data Transformation

To perform an accurate analysis, it is necessary to not use the data collected directly. Furthermore, the data needs to be cleaned and transformed into a usable format. This process involves replacing or eliminating missing values and discretizing certain continuous variables. Additionally, it entails removing redundant or unnecessary variables from the database before utilizing the available data, including removing null values and columns without unique values. The objective of this process is to improve the model's prediction rate with the highest accuracy.

Generally, the path generated from the car reverse test is not unified, as the number of points constituting it can vary due to the image processing results. Therefore, in this work, we propose a curve fitting technique to obtain a simplified and optimized path for further processing.

Curve fitting is primarily concerned with the challenge of approximating a parametric curve to a noisy set of data points. In the current research, the data points come from a single curve without any intersections. The tuning process begins with an initial curve and is then refined by minimizing an objective function that evaluates the quality of the fit and the smoothness of the curve. Figure 3 illustrates an example of the application result of curve fitting method on a pre-generated path.

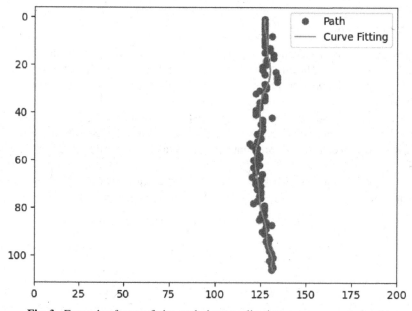

Fig. 3. Example of curve fitting technique application on pre-generated path

The next step is to select 50 data points from the resulting curve. Since the **y** coordinates are continuous values, they will be excluded from the dataset. Therefore, only the **x** values and the mean standard deviation between points will be retained for data mining process.

3.3 Data Mining Method

In this research, the data mining process is employed to predict and suggest the most accurate score for a driver's license applicant during the car reverse test. The study aims to investigate the impact of various Machine Learning classifiers (ML) algorithms for data analysis. Numerous classifiers have been discussed in the existing literature, including Decision Tree (DT), Support Vector Machine (SVM), Naïve Bayes, K Nearest Neighbors (KNN), and Multilayer Perceptron (MLP). Since no previous studies tackling the same subject could be found, this work will primarily focus on evaluating MLP, SVM, DT, and Naïve Bayes classifiers and examining their respective outcomes.

Support Vector Machine (SVM) Is a type of machine learning algorithm used for both classification and regression tasks. Its inception dates back to the proposal by Authors of [17], and since then, it has generated significant interest within the research community [18]. The SVM algorithm builds the hyper plane or a set of planes, which can effectively classify data into distinct classes. It relies on supervised learning, where experts control the training data, and it uses labelled training data to classify new, unseen data. SVM accomplishes this by mapping the input data into a higher-dimensional space, where it constructs a hyper plane that maximally separates the data. For multiclass classification problems, SVM decomposes them into multiple binary class problems, and it builds a combined multiclass SVM classifier using tools like LIBSVM and a practical guide for implementation as documented by Authors of [19] and [20].

Multilayer Perceptron (MLP) Are feed-forward networks consisting of simple processing units called neurons, which have at least one hidden layer. MLP classifiers are primarily used for classification tasks. The MLP network is composed of three layers: the input layer, hidden layer, and output layer. A neuron in the MLP network receives multiple inputs, denoted as C_1, C_2, C_3,...C_R, and each input is weighted by a corresponding element, denoted as $W_{1,1}$, $W_{1,2}$,...$W_{1,R}$. The weighted inputs are summed to give the net activation of the neuron, which is then passed through a transfer function to produce the neuron's output. Additionally, each neuron has a bias b, which is added to the weighted inputs to obtain the net input N [21]. The architecture of an MLP is shown in Fig. 4, where the net input N is computed as the sum of all weighted inputs and bias.

$$N = W_{1,1} + C_1 + W_{1,2} + C_2 + W_{1,3} + C_3 \ldots\ldots\ldots\ldots W_{1,R} + C_R + B$$

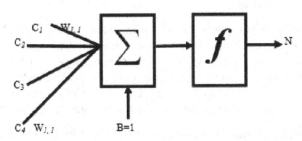

Fig. 4. MLP architecture review [21]

Naïve Bayes is a predictive modelling classifier that utilizes the Bayes Theorem to make probabilistic classifications. It assumes that features within a class are indifferent to presence of each other; in addition, this classifier is able to predict future possibilities in a class based on experiences. This classifier is suitable for both binary and multi-class classification problems.

For a uniform distribution, the formula for Naïve Bayes algorithm is presented as follows [15]:

$$P(c|x) = \frac{P(x|c).P(c)}{P(x)}$$

$$P(c|X) = P(x_1|c) \times P(x_2|c) \times \cdots \times P(x_n|c) \times P(c)$$

where:

P(c|x): the posterior probability of class (c, target) given predictor (x, attributes).
P(c): the prior probability of class.
P(x|c): the likelihood which is the probability of the predictor given class.
P(x): the prior probability of the predictor.

According to many researches such as [22], the Naïve Bayes classifier scores a high accuracy of prediction when applied in large databases.

Decision Trees (DT) Is a non-parametric technique widely used in decision analysis and operations research, which allows the analysis of small or large data structures and prediction of the target variable from other variables. DT classifiers are popular methods in data mining because they create a top-down tree structure using "if-then" rules to classify data. The branches of the tree represent different attributes and the leaf node represents a class. Decision trees are simple to use and do not require complex data representations, making them ideal for numeric and categorical data features. Algorithms ID3 and C4.5 are the main decision tree learning algorithms, which build decision trees using top-down greedy search method to check every attribute at every node of the tree. Building a decision tree involves two steps: building the tree and pruning the tree, with tree pruning used to correct any errors. Decision tree algorithm is simple, easy to analyse, very accurate and efficient in execution [23].

3.4 Performance Metrics

For any study, evaluating the performance of using machine learning classifiers or algorithms is an important task. The variation in results is justified depending on the different chosen evaluation metrics, along with the data-mining model that has been adopted. There are many types of evaluation metrics available to evaluate the efficiency and degree of success with respect to the given dataset. Therefore, this study focuses on using classification accuracy and F. Measure (F1 Score) which are defined as follows:

Accuracy: Is the percentage of correct predictions out of the total predictions made.

Accuracy = number of correct predictions/total number of predictions

Precision: Is the percentage of true positive results out of the total positive predictions.

$$Precision = True\ Positive/True\ Positive + False\ Positive$$

Recall: Is the number of true positive predictions divided by the sum of true positive and false negative predictions.

$$Recall = True\ Positive/True\ Positive + False\ Negative$$

F1 Score: Represent the Harmonic Mean between the precision and recall. The output of this metrics is ranged between 0 and 1. This metric confirms the accuracy of the given classifier with a greater value. Moreover, a higher precision with a lower recall gives a mostly accurate method.

$$F1\ Score = 2 * (Precision * Recall)/(Precision + Recall))$$

4   Results and Discussions

Our study involved testing the proposed classifiers, which included SVM, NB, and DT, using the Scikit-learn library. For MLP implementation, we utilized the Keras library. All tests were executed on a desktop machine with an Intel Core i5 1.80 GHz CPU and 8 GB of RAM. For software development, we used Python language under Windows 10 operating system.

To facilitate the development process, we employed a basic Multilayer Perceptron (MLP) with standard configurations. This involves incorporating one output layer that utilizes softmax as the activation function, which is commonly used for multiclass classification problem, MCXENT as the loss function, and Xavier as the weight initialization function. Additionally, we set the learning rate to 0.01.

For the remaining methods, default parameter values were applied. Gaussian Naïve Bayes was specifically chosen for its performance on small datasets. The linear kernel was utilized for SVM, and the 'gini' criterion was employed for node splitting in the case of DT.

All experiments are conducted with full training set. Therefore, we used the full dataset (50 record) for training. In addition, we deployed a set of 15 record size as a set for test validation.

Using performance metrics cited in section above, Table 3, illustrate the achieved outcomes of each studied classifier, specifically SVM, NB, DT and MLP.

By analyzing the results shown in Table 3, it is revealed that MLP classifier obtain about 93% of accuracy in classifying scores, which is higher compared to other methods. NB attained 88%, followed by SVM with an accuracy of 81%, while DT achieved a lower classification accuracy of 67%.

Table 3. Classification results using proposed machine-learning classifiers

Method	Accuracy (%)	Precision	Recall	F-Score
MLP	93	0.94	0.93	0.93
NB	88	0.88	0.8	0.8
SVM	80	0.85	0.8	0.81
DT	67	0.51	0.58	0.53

The comparison of results obtained by using studied classification techniques are presented in graphs shown in Fig. 5.

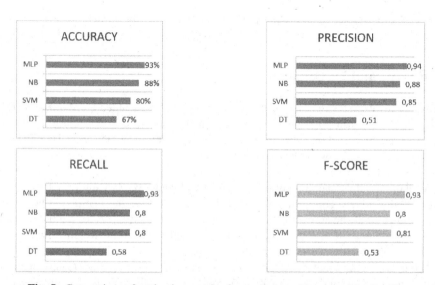

Fig. 5. Comparison of evaluation metrics for studied machine-learning classifiers

In order to predict the most accurate score for the test, the system calculates the percentage of appertaining of the given path to each known class. The class with highest percentage is then chosen. The complete results for each technique using five test samples are presented in Tables 4, 5, 6 and 7.

Among the five selected tests, our system's scores predictions matches the ones proposed by the supervisor, especially when using SVM and MLP classifiers, achieving a prediction accuracy of 100%. However, there is some differences observed when using NB (80%) or DT (40%).

In this study, we utilized a series of pre-generated paths obtained from tracking the vehicle during the car reverse test. It should be noted that the application of the curve fitting method might yield misleading outcomes for the classifier. This is primarily due to the implementation of image processing techniques in the preceding stages.

Table 4. Score prediction results using DT

Test	Scores						Ob-served score
	5	6	7	8	9	10	
Test 1	0%	0%	100%	0%	0%	0%	7
Test 2	100%	0%	0%	0%	0%	0%	6
Test 3	100%	0%	0%	0%	0%	0%	5
Test 4	0%	0%	100%	0%	0%	0%	10
Test 5	0%	0%	0%	0%	0%	100%	9

Table 5. Score prediction results using SVM

Test	Scores						Ob-served score
	5	6	7	8	9	10	
Test 1	4.71%	10.94%	31.55%	27.52%	14.13%	11.15%	7
Test 2	23.83%	48.38%	21.24%	2.64%	1.83%	2.07%	6
Test 3	42.14%	30.58%	17.06%	3.48%	2.67%	4.08%	5
Test 4	4.35%	6.18%	15.94%	29.77%	20.63%	23.12%	10
Test 5	2.60%	3.51%	7.29%	39.38%	26.03%	21.19%	9

The objective of this component of the system is to aid the supervisor in evaluating candidates who have successfully passed the car reverse test by suggesting the most accurate score. While this score is not mandatory for the supervisor's decision-making process, it should be reasonable and logical in order to improve the predictive capabilities of the system.

Table 6. Score prediction results using NB

Test	Scores						Ob-served score
	5	6	7	8	9	10	
Test 1	0%	0%	0%	100%	0%	0%	7
Test 2	0%	100%	0%	0%	0%	0%	6
Test 3	100%	0%	0%	0%	0%	0%	5
Test 4	0%	0%	3.42%	0%	0%	96.57%	10
Test 5	0%	0%	0%	0%	96.27%	3.73%	9

Table 7. Score prediction results using MLP

Test	Scores						Ob-served score
	5	6	7	8	9	10	
Test 1	0%	0%	100%	0%	0%	0%	7
Test 2	0%	100%	0%	0%	0%	0%	6
Test 3	98.70%	0%	1.3%	0%	0%	0%	5
Test 4	0%	0%	3.42%	0%	0%	100%	10
Test 5	0%	0%	0%	0%	100%	0%	9

5   Conclusion and Perspectives

In this paper, we employed data mining of driving license exam applicants to evaluate the performance of candidates during the car reverse test. The objective of this study was to analyze the potential influence of driving behavior during this test on the evaluation process and investigate the relationship between this behavior and learning outcomes. In this work, we used applicant's data collected after passing the test. Then, we have applied classification techniques using DT, NB, SVM and MLP classifiers in order to predict applicant's performance evaluation and propose it to the supervisor.

Generally, the obtained results showed that the prediction accuracy for MLP (93%) and NB (88%) are higher compared to SVM (80%) and DT (67%). Therefore, this disparity in favor of MLP leads to consider it as an efficient classifier to predict and evaluate the candidate's performance.

Our approach has shown promising results and has the potential to save time and effort, particularly by utilizing a single installed camera. The outcomes generated by the proposed program are expected to yield superior results compared to traditional methods that rely solely on human judgment.

For future work, we will respond to additional criteria stipulated by the law, particularly when the test requires termination and subsequent reevaluation by the applicant. To achieve this, we will integrate additional factors such as time and vehicle speed into the learning process. Moreover, we intend to assess the program's performance using real-time footage to compare the results with actual outcomes. Moreover, we will work on augmenting the number of participants on the experiment to facilitate a more comprehensive evaluation and provide the most accurate propositions.

References

1. Azi, A., Salhi, A., Jourhmane, M.: Line area monitoring using structural similarity index. International Journal of Advanced Computer Science and Applications **10**, 01 (2019)
2. Azi, A., Salhi, A., Jourhmane, M.: Car tracking technique for DLES project. In: Fakir, M., Baslam, M., El Ayachi, R. (eds.) Business Intelligence: 7th International Conference, CBI 2022, Khouribga, Morocco, May 26–28, 2022, Proceedings, pp. 279–293. Springer International Publishing, Cham (2022). https://doi.org/10.1007/978-3-031-06458-6_23
3. Zheng, X., Yang, P., Duan, D., Cheng, X., Yang, L.: Real-time driving style classification based on short-term observations. IET Commun. **16**(12), 1393–1402 (2022). https://doi.org/10.1049/cmu2.12405
4. Leite da Silva, C., May Petry, L., Bogorny, V.: A survey and comparison of trajectory classification methods. In: 2019 8th Brazilian Conference on Intelligent Systems (BRACIS). IEEE (2019). https://doi.org/10.1109/bracis.2019.00141
5. Xue, Q., Wang, Ke., Jian John, Lu., Liu, Y.: Rapid driving style recognition in car-following using machine learning and vehicle trajectory data. J. Adv. Transp. **2019**, 1–11 (2019). https://doi.org/10.1155/2019/9085238
6. Lee, J.-G., Han, J., Li, X., Gonzalez, H.: Traclass: trajectory classification using hierarchical region-based and trajectory-based clustering. Proc. VLDB Endowment **1**(1), 1081–1094 (2008)
7. Bolbol, A., Cheng, T., Tsapakis, I., Haworth, J.: Inferring hybrid transportation modes from sparse GPS data using a moving window SVM classification. Comput. Environ. Urban Syst. **36**(6), 526–537 (2012)
8. Lee, J.-G., Han, J., Li, X., Cheng, H.: Mining discriminative patterns for classifying trajectories on road networks. IEEE Trans. Knowl. Data Eng. **23**(5), 713–726 (2011)
9. Dabiri, S., Heaslip, K.: Inferring transportation modes from GPS trajectories using a convolutional neural network. Transp. Res. Part C Emerg. Technol. **86**, 360–371 (2018)
10. Tragopoulou, S., Varlamis, I., Eirinaki, M.: Classification of movement data concerning user's activity recognition via mobile phones. In: Proceedings of the 4th International Conference on Web Intelligence, Mining and Semantics (WIMS14), p. 42. ACM (2014)
11. Varlamis, I.: Evolutionary data sampling for user movement classification. In: 2015 Evolutionary Computation, pp. 730–737. IEEE (2015)
12. Xiao, Z., Wang, Y., Fu, K., Wu, F.: Identifying different transportation modes from trajectory data using tree-based ensemble classifiers. ISPRS Int. J. Geo Inf. **6**(2), 57 (2017)

13. Etemad, M., Júnior, A.S., Matwin, S.: Predicting transportation modes of GPS trajectories using feature engineering and noise removal. In: Bagheri, E., Cheung, J.C.K. (eds.) Advances in Artificial Intelligence, pp. 259–264. Springer International Publishing, Cham (2018). https://doi.org/10.1007/978-3-319-89656-4_24

14. Romero, C., Ventura, S.: Educational data mining: a survey from 1995 to 2005. Expert Syst. Appl. **33**(1), 135–146 (2007). https://doi.org/10.1016/j.eswa.2006.04.005

15. Sultana, J., Farhat, N., Kazi, S., Usha, M.: Design engineering an SPPML framework to classify students performance using MLP. **2021**, 14235–14244 (2022)

16. Romero, C., Ventura, S., Garcia, E.: Data mining in course management systems: moodle case study and tutorial. Comput. Educ. **51**(1), 368–384 (2008). https://doi.org/10.1016/j.compedu.2007.05.016

17. Boser, B., Guyon, I., Vapnik, V.: A training algorithm for optimal margin classifiers. In: Proceedings of the Fifth Annual Workshop on Computational Learning Theory, Pittsburgh (1992)

18. Srivastava, D.K., Bhambhu, L.: Data classification using support vector machine. Int. J. Theor. Appl. Inf. Technol. **12**(1), 1–7 (2009)

19. Jindal, A., Dhir, R., Rani, R.: Diagonal features and SVM classifier for handwritten Gurumukhi character recognition. Int. J. Adv. Res. Comput. Sci. Softw. Eng. **2**(5), 505–508 (2012)

20. Hsu, C.-W., Chang, C.-C., Lin, C.-J.: A Practical Guide to Support Vector Classification

21. Nazzal, J.M., EL-Emary, I.M., Najim, S.A.: Multilayer preceptron neural network (MLPs) for analyzing the properties of Jordan oil shale. World Appl. Sci. J. **5**(5), 546-552 (2008)

22. Shedriko, R.E., Husein, I., Zufria, I., Nasution, H.: Naive Bayes algorithm to predict student success in a course. IJAST **29**(08), 2964–2974 (2020)

23. Topîrceanu, A., Grosseck, G.: Decision tree learning used for the classification of student archetypes in online courses. Procedia Comput. Sci. **112**, 51–60 (2017)

Optimization and Decision Support

A Comparative Study on the Implementation of Blockchain in Supply Chain Models

Dhanashri Joshi⬤, Atharva Naikwadi⁽✉⁾⬤, Rohit Mokashi⬤, Mohit Pande⬤,
and Sitanderpal Singh⬤

JSPM Rajarshi Shahu College of Engineering, Pune, India
dmjoshi_comp@jspmrscoe.edu.in, atharva.s.naikwadi@gmail.com

Abstract. This research intends to investigate the existing state, prospective uses in supply chain, future scope of this technology, dangers, and obstacles that we may encounter. We discuss the importance of blockchain in supply chain in Indian context. Each customer should be assured of the quality and credibility of items they would be using. This technology is most known for its application in crypto currency. It has the potential of increasing its reach to much more domains. The blockchain network is comparatively a better option than a traditional approach as each new block of data is attached to the existing blockchain in a linear and chronological way. With each block connected to the preceding one, any bad actors wishing to tamper with data would need to decode prior blocks before reaching their desired data and rewriting the chain ahead while manipulating the rest of the nodes in a similar approach. As tampering with the data is both technologically difficult and commercially unfeasible, the security of a blockchain becomes two-fold.

Keywords: Blockchain · Solidity · Ethereum · Smart Contracts · Ethereum Virtual Machine

1 Introduction

Blockchain technology was first proposed as a prototype in 1982 by David Chaum. Although the first real life application that is Bitcoin was bought into existence in 2009 by Satoshi Nakamoto. We live in Industry 4.0 which is based on sharing on information. Now, the requirement of faster and much secure way of providing and storing information is more than ever. Blockchain provides a platform which all users can access is transparent and is highly secured. The most important element of a blockchain network is smart contract. A smart contract can be considered as a digital bond between two parties which defines the terms and conditions for secured trade.

A blockchain comprise of blocks which store the necessary information of the product like its timestamp and condition in a digital format electronically. These blocks are connected to one another. Each block has the data, the hash of its own and hash of previous block. Hence, making it one of the most successful and secure networks.

R. El Ayachi et al. (Eds.): CBI 2023, LNBIP 484, pp. 105–116, 2023.
https://doi.org/10.1007/978-3-031-37872-0_8

There are many types of a blockhain, such as: public blockchain network, private blockchain network, permissioned blockchain network, consortium blockchain network.

Blockchain plays a major role in many applications such as: money transfers, insurance, elections, Non Fungible Token (NFTs), data storage, data security, supply chain monitoring, protection of intellectual property rights like royalty and copyrights, etc.

Blockchain is projected to broaden its application scope to include the Internet of Things (IoT), significant data analysis, law-making / enforcement, and finance. Blockchain technology has the potential to drastically alter how we live and work in the future. The Global Blockchain Market is estimated to reach USD 34 billion by 2026, with a 45 percent growth rate.

2   Advantages of Blockchain

- Transparency- All members have the responsibility to update their data about the product. A collection of accurate data improves credibility and trust amongst the members or users. For example: Suppose a crate of mangoes is going to be transferred from the farm to the store, this technology will keep timestamp of the start of the travel of the product from the farm to the warehouse till the end destination.
- Security- This technology is based upon distributed ledger concept. This implies there is no single owner for the system or network. The network consists of multiple users and it is thus very difficult for one user to gain a stake for making decisions. Thus, the data in blockchain technology is not possible to change or tampered with. Also, a blockchain has multilayer protection that ensures safety from hackers.
- Analytics- This technology is not just about being a secured storage system but it is also capable of giving solutions to the complicated data problems that are updated. It can create, maintain and analyse necessary statistics that help in improving the user interface.
- Improves Coordination- One of the most important advantages of blockchain is that it can help improve coordination between different components of a supply chain. It improves the scope of data sharing which is an important aspect.
- Open source- The framework we would be using is Ethereum. The programming language used in the framework would be Solidity. The Ethereum framework is an open-source distributed ledger.
- Efficiency- It is easy to identify a defective product or miscounting, because the life cycle of a product is maintained at every level.
- Reduces counter fitting and duplication- The use of blockchain will ensure that counterfeit and fake products are not bought by the customers and thus ensures safety of users.

3   Disadvantages

- High upfront cost- The cost for deployment can be high in real world for larger implementation. The newer or upcoming firms might find it difficult to develop their own blockchain whereas the bigger ones can make it easily. There are expenses accustomed with employing network blockchain developers, which comparatively cost more than traditional developers due to their specialization and expertise. Also, there are costs for maintenance, licensing, documentation and planning.

- Relies on users- It relies on the network or ledger. Number of users and their active time is directly proportional to the value of that network. In order for it to succeed, all companions of the network have to use the platform.
- Consumption of Energy- The proof-of-work technique is extremely energy-intensive, requiring miners to use a significant amount of computational power to accomplish the computations.
- Impact on nature- As it uses a lot of energy; it is obvious that it would leave its footprint on nature. Thus has a bad impact on environment.

4   Working of Blockchain

4.1   Components of a Blockchain

Blockchain consists of blocks, nodes and miners.

Blocks-Every chain comprises of several blocks, each block in the chain has three basic elements, the data stored inside the block. When a block is created, a nonce (a 32-bit whole number) is generated at random, and a hash is created for that block. The nonce is than incorporated with a 256-bit integer code and is incredibly small.

Miners-Mining is the technique through which miners add new blocks to the existing blockchain. Each and every block in the blockchain has its own unique identification and hash, and it also gives reference to the previous block in the, making mining of blocks difficult on large blockchains. Miners make use of special software to generate unique and valid hash for every block in the chain using nonce. Because the nonce and hash are only 32 bit and 256 bit respectively and there are nearly four billion combinations of nonce and hash, which need to be searched before finding the appropriate one. This means that miners have discovered a "golden nonce" and the block is added to the end of the chain.

Nodes-One of the most important advantages of blockchain is that it is decentralized. A single machine or person cannot be the master of the chain. Instead, a distributed ledger is formed by the nodes that connect to the chain. A node is any form of technological device that maintains replicas of the chain and keeps the system operating.

4.2   What is Solidity Programming and Its Working

The Ethereum Network team created Solidity, an object-oriented programming language developed specifically for designing and implementing smart contracts on Blockchain platforms. It's utilized in the blockchain system to create smart contracts that apply business logic and build a chain of transactional data. It's used to develop and build machine-level programs on the Ethereum Virtual Machine (EVM). In many aspects, it is similar to C and C++, and it is simple to learn and grasp. For example, main function in C is same as contract in Solidity. Variables, functions, classes, arithmetic operations, string manipulation, and plenty of other concepts are all present in Solidity programming, as they are in other languages.

Each node has its own copy, and the system algorithmically approves each and every block that is recently mined in order to update the blockchain and make it trustworthy and valid also. As blockchains are transparent, all the action can be easily inspected and investigated. All the users of the blockchain are given a unique alphanumeric id code in order to track their actions.

4.3 Ethereum and Its Working

Ethereum is based on a blockchain network. A blockchain is a formation of blocks containing data. Every block comprises of the data, previous hash and current hash. It's distributed in the sense that everyone on the Ethereum network has an identical copy of the ledger, which allows them to observe all previous transactions. It's decentralized in the sense that the network isn't run or maintained by a single organization, but rather by all of the distributed ledger owners. Blockchain transactions employ cryptography to keep the network secure and validate transactions. As an incentive, participants are given cryptocurrency tokens. In the Ethereum system, these tokens are known as Ether (ETH). Ether can also be used as money for. Its value has risen significantly in past few years, which makes it an excellent investment. Many developers may develop apps that execute in the Ethereum the same way it executes on a computer, which makes Ethereum unique.

4.4 The Ethereum Virtual Machine (EVM)

The Ethereum Virtual Machine (EVM) is a system that uses the Ethereum blockchain to run (EVM).

The EVM is Ethereum's processor and memory platform, which lets developers to construct and interact with clean smart contracts. To create smart contracts, Ethereum developers use Solidity, a programming language similar to JavaScript and C++. These Solidity smart contracts can be examined by humans, but not by robots. As a result, it must be converted into EVM-friendly low-level device instructions (also known as opcodes). It's critical to understand that each Ethereum node has its own EVM.

Every node processes the smart agreement and transaction using their own Ethereum Virtual Machine when a customer makes a transaction to an Ethereum clever agreement (EVM). Each node may also see what the ultimate outcome will be in this simulated environment, as well as whether or not the final results will result in a legitimate transaction or not. If all nodes reach the same lawful conclusion, changes are done and the amended Ethereum country is recorded on the blockchain.

4.5 Smart Contracts

Smart contracts are high-level programming programs that can be compiled into EVM and then published to the Ethereum network.

A code that is created to automatically execute functions based on the terms and conditions decided upon by network stakeholders. It works as a virtual agent that verifies transactions without the need for third-party intervention. Our proposed approach consists of four smart contracts, each of which is focused on a specific task. The registration contract focuses on registering all network members, while the inventory level contract specifies the amount of inventory left for each supplier as well as the product descriptions offered by suppliers. The order management contract deals with handling and managing orders, and the reputation contract ranks vendors based on their product quality, distribution efficiency, and honesty.

5 System Architecture

Figure 1.

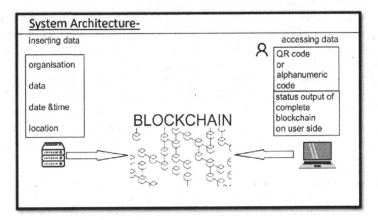

Fig. 1. System Architecture

6 Project Methodology

Figure 2.

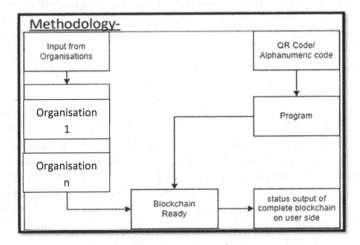

Fig. 2. Methodology

7 Challenges

- Lack of experience- Our team is new to working with this technology. We have to learn the working of Ethereum framework and Solidity programming language. Thus, lack of experience can be a major challenge for us.
- Fast changes- In the modern world, there are constant changes that are made every now and then. It is important for us to be aware of latest changes in the market related to the blockchain technology. Maybe, newer tools, platforms or programming language can be invented in given time frame.
- Lack of coordination- We might not be able to coordinate together or agree on sometimes, there might be lack of communication within the group.
- Public Awareness: Through many investors are started investing in blockchain still there's along way to go for the blockchains to be fully adopted by masses. Still, there are people in the world unaware about usage, features and many more advantages of blockchain.

8 Future Scope

The use of blockchain will continue to impact business operations. Blockchain facilitates product traceability and fosters trust between partners (producers and end users). Due to COVID-19, people came to know that health is everything, and everyone knows our health is directly related to the food that we consume. The demand is for the replacement of the traditional supply chain with no security, transparency, and traceability of products. With devices that work with Blockchain, customers will be able to track their product life cycle in the supply chain. Block-chain is an efficient method with low maintenance and high output. A collection of accurate data improves credibility and trust amongst the members or users. Thus, the data in blockchain technology is not possible to change or tamper with. The most important advantage is that it can help improve coordination between different components of a supply chain. It improves the scope for exchange of data which is important.

9 A Technical Overview of the Project

This project is capable of keeping track of the products over the complete path from producer to consumer. Technologies used for this project are some common blockchain technologies like solidity programming language for writing the smart contract that build up the base of the blockchain and control it's flow. As a blockchain product once uploaded to the main blockchain or main net cannot be rollbacked, we are using ganache for the purpose of making and testing blockchain or testnet on which we can run our project, test it and make necessary changes on the way. Writing solidity smart contracts, compiling them, and pushing it to the testnet is a tedious task in itself. So, to tackle these problems we are taking in use of the Remix IDE which is one of the best IDEs or integrated development environment for solidity and Ethereum. Remix IDE provides code compiler and other features like it automatically connects to the mainnet via the IP address and port number on which our test net is running, it can also help us deploy the code directly on mainnet of Ethereum, and it also provides some initial testnets for testing so that we don't have to setup a new test network.

10 Methodology

Our project would consider the scenario in context of the Indian market. Like, consider a small scale fruit vendor: if a customer asks for assurance of quality whether the given product is organic and how long would it remain fresh then this project would turn useful. The project would consist of a system comprising of a blockchain network made of Ethereum framework, an alphanumeric code or a QR scanner and an user interface like a website. Once the customer scan the QR code on the food packet or any consumables, they would be redirected to a site which mentions the real time data of the product in the supply chain. It would mention all the neccessary information like the origin of produce, the date of yield, the locations of warehouses it was stored and much more. This information cannot be tampered with due to the usage of the blockchain which helps by providing two steps of security. First the network is decentralised and to make any change, more than half of the network users should approve it. Second, the hashing mechanism which stores the hash value of the previous block helps detect temperament. Making changes in data would result in change in the hash value of the block which in turn would affect hash values of all the next blocks.

11 Discussion

This technology has been used in various other industries and retail chains around the world, but this project considers the Indian market context where the target is to make the untampered information available to the customer even at the local levels at an affordable cost.

12 Conclusion

In conclusion, this research paper has explored the significance of blockchain technology in the context of the supply chain, particularly within the Indian context. By investigating its current state, prospective uses, and future scope, as well as the potential dangers and obstacles associated with its implementa-tion, we have highlighted the transformative potential of blockchain technol-ogy.

Blockchain's ability to ensure the quality and credibility of items for each customer is a key aspect of supply chain management. Traditionally, trust in the supply chain has been a challenge, but blockchain offers a promising so-lution. Moreover, while blockchain is widely recognized for its application in cryptocurrency, its reach extends far beyond that domain, presenting oppor-tunities for various industries.

Compared to traditional approaches, the blockchain network provides a supe-rior option due to its linear and chronological structure. Each new block of data is securely attached to the existing blockchain, making it technologically difficult and commercially unfeasible for bad actors to tamper with the data. This inherent security feature enhances the trustworthiness and reliability of the supply chain system.

In light of these findings, it is evident that blockchain technology has the po-tential to revolutionize supply chain management in India and beyond. How-ever, successful implementation will require overcoming challenges such as scalability, interoperability, and regulatory frameworks.

As further research and advancements in blockchain technology continue to unfold, it is recommended that businesses and policymakers in the supply chain sector actively explore and adopt blockchain solutions. By leveraging the secure and transparent nature of blockchain, organizations can enhance efficiency, traceability, and customer satisfaction.

In conclusion, the integration of blockchain technology in the supply chain holds immense promise and presents exciting opportunities for businesses, consumers, and the overall economy. Embracing blockchain can pave the way for a more secure, efficient, and transparent supply chain ecosystem, ulti-mately benefiting all stakeholders involved.

13 Literature Survey of Base Paper

Table 1

Table 1. Reference Summary of Base Papers

Year & Method	Output achieved & Summary	Future scope	Remark
2020 - When Blockchain Meets Supply Chain [1]	In this research study, we re-evaluated current blockchain investigations in the context of SCM. Several studies have looked into how blockchain technology might impact conventional supply chains. A transaction cost viewpoint was employed in several studies. Blockchain, being a decentralized network, has the potential to minimise user trust problems	In the future, other blockchain-related technological challenges, such as intransigence, privacy, reliability, and scalability, will need to be tackled. There are few such endeavors, and quantitative research in these fields is still scarce	This research will provide an understanding of recent blockchain-related sustainable supply chains, as well as outline research constraints and potential research objectives

(continued)

Table 1. (*continued*)

Year & Method	Output achieved & Summary	Future scope	Remark
2019 - Incorporating Block Chain Technology in Food Supply Chain (Taken in Indian context) [2]	With the aid of block chain and the numerous benefits it provides, a block chain model in the context of food supply chain is appropriate	This paper's future prospects include assisting in the development of a blockchain supply chain model	The study looked at the areas of block chain, supply chain, and difficulties in the food supply that may be addressed by using a block chain model
2020 - Enhancing Vendor Managed Inventory Supply Chain Operations Using Blockchain Smart Contracts [3]	A blockchain on the concept of VMI system can improve quality and assure supply chain performance by facilitating communication between different trade sections	We intend to create and build decentralized packages in the future for the users	We wish to fulfil the diverse requirements of the supply chain in the context of the VMI process
2020 - Blockchain-Based Safety Management System for the Grain Supply Chain [4]	This paper aims to improve the management for grain sector and also help in the distribution network that might help to improve the accuracy and traceability	Future research may merge IoT technology with blockchain to assure genuine collection of data	There are currently several flaws in logistics management, the most important of which is lack of reliable data collection. More study on how to assure the reliability of information sources is required
2020 - How Blockchain Enhances Supply Chain Management: A Survey [5]	Many issues have been found. Various techniques and frameworks, as well as answers to difficulties faced, are explored	Blockchain can be used to safely and effectively move user data throughout systems and platforms. The program could be used to keep track of ownership, patents, and other information	Blockchain provides a viable messaging service with vendor lock assurances for the establishment of powerful and expense decentralized supply chain management

(*continued*)

Table 1. (*continued*)

Year & Method	Output achieved & Summary	Future scope	Remark
2019 - A Survey on Using Blockchain in Trade Supply Chain Solutions [6]	Blockchain technology has the ability to streamline the monitoring system and secure the integrity of shared data to improve the commercial distribution network. This technology's data integrity and transparency features make it perfect for selling highly regulated items like medicines	As a consequence, implementing a blockchain-based information-exchange system is certain to improve the customs authority's operational technique and facilitate worldwide commerce	Inside this distribution network, block chain has substantial opportunities
2022 - Supply Chain Inventory Sharing Using Ethereum Blockchain and Smart Contracts [7]	Information exchange is very crucial for a financially successful distribution network and this paper explores its importance	Using system design, methods, flowchart, and tests scenarios, the recommended solution may be adjusted to match a range of product categories in the distribution network	Ethereum is a decentralised, open-source blockchain technology that will be helpful in this market, but there are limited resources and study on it
[2021] Design of Supply Chain System Based on Blockchain Technology [8]	We demonstrated a blockchain-based supply chain network and tried to fix its problems	The system can meet the supply chain's basic requirements, but it needs more research in terms of bandwidth and safety	Blockchain is a decentralised, tamper-resistant digital ledger built on a variety of technologies. It has a reliable database, as well as confidentiality and dependability

(*continued*)

Table 1. (*continued*)

Year & Method	Output achieved & Summary	Future scope	Remark
2020 - A Content-Analysis Based Literature Review in Blockchain Adoption within Food Supply Chain [9]	This paper was one of the pioneers to examine the use of blockchain in the food supply chain. It provides an understanding on the topic of blockchain and will help the not only the professionals but also the corporate decision makers	This research brings together 24 studies on the food system but does not name individual products. Future research might focus on specific meals and provide more precise results	We'd like to call attention to a few restrictions and possible research subjects. To begin with, the research is based on a review of 24 papers that may not be enough to eliminate study biases

Bibliography

1. Chang, S.E., Chen, Y.: When blockchain meets supply chain: a systematic literature review on current development and potential applications. https://doi.org/10.1109/ACCESS.2020.2983601
2. Omar, I.A., Jayaraman, R., Debe, M.S., Hasan, H.R., Salah, K., Omar, M.: Supply chain inventory sharing using ethereum blockchain and smart contracts. https://doi.org/10.1109/ACCESS.2021.3139829
3. Gonczol, P., Katsikouli, P., Herskind, L., Dragoni, N.: Blockchain implementations and use cases for supply chains-a survey. https://doi.org/10.1109/ACCESS.2020.2964880
4. Shakhbulatov, D. (Member, IEEE), Medina, J. (Graduate Student Member, IEEE), Dong, Z.: How blockchain enhances supply chain management: a survey. https://doi.org/10.1016/j.patrec.2020.04.009
5. Duan, J., Zhang, C., Gong, Y., Brown, S., Li, Z.: A content-analysis based literature review in blockchain adoption within food supply chain. https://doi.org/10.3390/ijerph17051784
6. Zhang, J., Zhong, S., Wang, T., Chao, H.-C., Wang, J.: Blockchain-based systems and applications: a survey. https://jit.ndhu.edu.tw/article/view/2217
7. Ray, P., Harsh, H.O., Daniel, A., Ray, A.: Incorporating block chain technology in food supply chain. https://doi.org/10.18843/ijms/v6i1(5)/13
8. Enhancing vendor managed inventory supply chain operations using blockchain smart contracts. https://doi.org/10.1109/ACCESS.2020.3028031
9. Zhang, X., et al.: Blockchain-based safety management system for the grain supply chain. https://doi.org/10.1109/ACCESS.2020.2975415
10. Li, J., Song, Y.: Design of supply chain system based on blockchain technology. https://doi.org/10.3390/app11209744
11. Kushwaha, S.S., Joshi, S., Singh, D., Kaur, M., Lee, H.-N.: Systematic Review of Security Vulnerabilities in Ethereum Blockchain Smart Contract. https://doi.org/10.1109/ACCESS.2021.3140091
12. Hu, T., Li, B., Pan, Z., Qian, C.: Detect defects of solidity smart contract based on the knowledge graph. https://doi.org/10.1109/TR.2023.3233999

13. Zheng, X.R., Lu, Y.: Blockchain technology – recent research and future trend. https://doi.org/10.1080/17517575.2021.1939895
14. Sangari, M.S., Mashatan, A.: A data-driven, comparative review of the academic literature and news media on blockchain-enabled supply chain management: Trends, gaps, and research needs. https://doi.org/10.1016/j.compind.2022.103769
15. Chang, A., El-Rayes, N., Shi, J.: Blockchain technology for supply chain management: a comprehensive review. https://doi.org/10.3390/fintech1020015

A Hybrid Approach of Dijkstra's Algorithm and A* Search, with an Optional Adaptive Threshold Heuristic

Lhoussaine Ait Ben Mouh[1]([⊠])[iD], Mohamed Ouhda[2][iD],
Youssef El Mourabit[1][iD], and Mohamed Baslam[1][iD]

[1] Moulay Slimane University, FST Beni Mellal, Beni-Mellal, Morocco
aitbenmouh@yahoo.fr, {y.elmourabit,m.baslam}@usms.ma
[2] Moulay Slimane University, EST Khenifra, Khenifra, Morocco

Abstract. This study is a part of the trajectory planning applied to harvest system work where mobile robots must be able to navigate safely the environment to look for palmer crops. Many constraints can be faced, such as crop selection as maturity changes over time, searching for the most mature palmer, avoiding different kinds of obstacles, robot speed control, and the cost of moving from an initial point to a goal target. After studying different trajectory planning approaches and their applications [8], we conclude that some of these methods can be combined to design a new, powerful approach based on the accurate property of Dijkstra and the heuristic function of A Star.Dijkstra is known as a powerful algorithm based on graph mapping and reducing the path cost, and A Star on the other side is one of the best guides for path searching due to the heuristic function that avoids exploring all environment nodes and only those leading to the goal. Combining Dijkstra's algorithm and the A* (A-star) algorithm can lead to a more efficient pathfinding approach. Dijkstra's algorithm [4] is a well-known method for finding the shortest path between two nodes in a graph, while the A* algorithm is an extension of Dijkstra's algorithm that uses heuristic estimates to guide the search towards the goal node. By combining these two algorithms, we can use Dijkstra's algorithm to explore the graph and generate a good initial estimate of the path cost, then use the A* algorithm to refine the estimate and guide the search towards the goal node. This paper explores the utilization of trajectory planning in a harvesting system. By employing both the Dijkstra and A* algorithms, we propose a hybrid approach to ensure optimal timing for finding a path. We conduct a comparative analysis to evaluate the performance of the new approach by comparing the application of a single algorithm versus the hybrid approach across various graph sizes.

Keywords: Trajectory planning · Path Planning · Mobile Robot · Artificial Intelligence

© The Author(s), under exclusive license to Springer Nature Switzerland AG 2023
R. El Ayachi et al. (Eds.): CBI 2023, LNBIP 484, pp. 117–133, 2023.
https://doi.org/10.1007/978-3-031-37872-0_9

1   Introduction

As a part of Trajectory Planning Applied to the harvesting system studies, it is so interesting to study the optimisation problem of find the optimal path between a starting point and goal. Dijkstra is one of the best algorithm on searching and finding the shortest path in a given Graph, but in our case we have to manage to work with large Graph and find quickly as possible a path in an acceptable timing. A star on the other side is very efficient approach on finding a path. This approach combines the strengths of different pathfinding algorithms to produce a more efficient and effective pathfinding solution [5]. One common approach is to combine Dijkstra's algorithm with the A* algorithm. Dijkstra's algorithm is used to explore the graph and build a search tree, while the A* algorithm is used to guide the search towards the goal node. By combining these two algorithms, we can leverage the optimality and completeness of Dijkstra's algorithm with the efficiency of the A* algorithm. In a hybrid pathfinding approach, the search space is divided into two regions, a Dijkstra region and an A* region. The Dijkstra region covers the nodes that are closer to the start node, and the A* region covers the nodes that are closer to the goal node. The boundary between these two regions is dynamically adjusted during the search process [17]. The hybrid approach is particularly useful for large, complex search spaces, where the A* algorithm alone may not be sufficient, or Dijkstra's algorithm may be too slow. By using a hybrid approach, we can find a path efficiently without sacrificing the optimality or completeness of the solution. In this study, we delve into the implementation of trajectory planning within a harvesting system. Our approach involves incorporating both the Dijkstra and A* algorithms to create a hybrid method that optimizes path-finding timing. Through a comparative analysis, we assess the performance of this novel approach by contrasting it with the use of a single algorithm, considering different graph sizes.

2   Related Works

Hybrid approaches in trajectory planning offer innovative solutions by combining different algorithms or techniques to enhance efficiency, optimality, adaptability, and robustness. Within the domain of hybrid approaches, several notable combinations have been explored. One set of approaches combines the A* algorithm with other methods. A* RRT, for instance, merges the exploration capabilities of Rapidly-exploring Random Trees (RRT) with the guidance provided by A*, enabling efficient trajectory planning in complex environments [3].

Another hybrid approach, A* D* (Dynamic A*), integrates the A* algorithm with D* to dynamically adapt trajectories based on changes in the environment or obstacles, ensuring flexibility and responsiveness. D* Lite, a variant of D* (Dynamic SWSF-FP), has also been involved in hybrid trajectory planning.

D* Lite RRT combines D* Lite with the RRT, leveraging the adaptability of D* Lite to improve efficiency and optimality in trajectory planning, particularly in dynamic environments. Additionally, D* Lite PRM (Probabilistic Roadmaps) combines D* Lite with PRM to efficiently generate collision-free trajectories, particularly in high-dimensional spaces [13].

These hybrid approaches in trajectory planning offer promising avenues for improved trajectory generation, considering the unique strengths of each constituent algorithm or technique. By leveraging the combined power of multiple methods, these hybrid approaches provide innovative solutions to the challenges of trajectory planning in various scenarios.

3 Materials and Methods

3.1 Dijkstra Algorithm

Dijkstra is an algorithm that can determine the shortest routes between nodes in a graph. It can be utilized for both directed and undirected graphs, provided that all edges have non-negative weights [16]. The algorithm guarantees that it will find the shortest path between two nodes as long as all edges have non-negative weights. To use Dijkstra's algorithm, we begin by selecting a starting node and assigning it a distance of 0. We then add this node to a priority queue and continue to iterate until we have visited all nodes in the graph. In each iteration, we extract the node with the smallest distance from the priority queue and examine its neighbors. For each neighbor, we calculate the distance to that node by adding the weight of the edge between the two nodes to the distance of the current node. If this distance is less than the current distance assigned to the neighbor, we update the neighbor's distance and add it to the priority queue. The algorithm terminates when we have visited all nodes in the graph, and the distance assigned to the destination node will be the shortest path from the starting node to that node. To reconstruct the path, we can backtrack from the destination node to the starting node, following the nodes with the smallest distances at each step [10]. By following a simple and intuitive set of steps, we can efficiently compute the shortest path between any two nodes in a connected graph with non-negative edge weights.

3.2 A Star Algorithm

The A* algorithm means of finding the shortest path between an origin and destination node in a graph. It utilizes an estimation of the distance from any given node to the destination node, which is represented by the function $h(i)$. One common example of $h(i)$ is the Euclidean distance measure. To begin the algorithm, each node is assigned a distance value $d(i)$ of positive infinity except for the origin node, which is assigned a value of 0. The set S is initialized to only contain the origin node. In each iteration of the algorithm, the node with the lowest distance value in set S is selected and removed from S. If this node is the destination node, the algorithm terminates. Otherwise, the algorithm updates the distance values of all neighboring nodes of the selected node and adds them to S if they are not already present. The distance value for a neighboring node is calculated by adding the weight of the arc connecting the nodes to the distance value of the selected node and the estimated distance from the neighboring node to the destination node. The estimated distance is calculated using the function $h(i)$. The A* algorithm is similar to the Dijkstra algorithm in that both maintain a set of candidate nodes and use a best-first method to select the next node to expand. However, A* differs from Dijkstra in that it includes a forward-looking component in its distance function, which estimates the length of the shortest path to the destination from a specific node. This component, $h(i)$, helps to direct the search space towards the destination [15].

3.3   Hybrid Approach

The hybrid approach of combining Dijkstra's algorithm with A*'s algorithm proove that it's more efficient than using only A* or Dijkstra's algorithm in situations where the search space is complex and the goal is far from the start node. For example, consider a scenario where a robot needs to navigate a maze-like environment with multiple obstacles to reach a goal location. In this case, the A* algorithm alone may not be efficient enough to find the optimal path, as it has to explore a large number of nodes to reach the goal. Dijkstra's algorithm can find the optimal path, but it can take a long time to explore the entire graph, especially if the graph is large and complex [16]. However, by using a hybrid approach, we can leverage the strengths of both algorithms. The initial search can be done using Dijkstra's algorithm, which will explore the nodes closest to the start node and build a search tree Algorithm 2. Once the algorithm reaches the boundary of the A* region, it can switch to the A* algorithm to guide the search towards the goal node. This approach allows the algorithm to explore the most promising nodes near the start node first, ensuring that the robot can begin moving towards the goal location as quickly as possible. At the same time, it ensures that the algorithm does not waste time exploring nodes that are far from the goal location. Combining Dijkstra's algorithm with A*'s algorithm is more efficient than using only A* or Dijkstra's algorithm in scenarios where the search space is complex, and the goal is far from the start node [18]. A mobile robot harvesting ripe palm fruits from a plantation and explore why using a hybrid approach combining Dijkstra's algorithm and A* algorithm is advantageous compared to using either algorithm alone:

1. Dijkstra's Algorithm Alone:
   - If we were to use Dijkstra's algorithm alone, the robot could efficiently map the plantation by considering the distances between palm trees. However, Dijkstra's algorithm does not take into account specific information about the ripeness of the fruits.

- As a result, the robot may follow paths that are optimal in terms of distance but may not necessarily lead to a higher yield of ripe fruits. It might spend unnecessary time reaching trees with unripe fruits, or miss out on areas with a higher concentration of ripe fruits.

2. A* Algorithm Alone:

- If we were to use the A* algorithm alone, the robot would consider both distance and fruit ripeness. It could plan its path based on the estimated yield of ripe fruits, prioritizing areas with a higher concentration of ripe fruits.
- However, without an initial map created by Dijkstra's algorithm, the robot may have difficulty efficiently navigating the plantation. It would have limited knowledge of the spatial layout of the palm trees, potential obstacles, or efficient routes between trees.

There many Advantages of the Hybrid Approach:

1. Efficient Mapping:
   Dijkstra's algorithm enables the robot to create a map of the plantation, considering distances and obstacles. This initial mapping provides valuable information about the spatial layout of the palm trees, which is crucial for effective navigation [2].
2. Optimal Harvesting Path:
   The hybrid approach permits to the robot can navigate efficiently based on the map while prioritizing areas with a higher concentration of ripe fruits.
3. Improved Efficiency:
   The hybrid approach helps the robot avoid unnecessary detours to unripe fruit trees, saving time and energy. It also ensures that the robot focuses on areas with a higher likelihood of yielding a greater quantity of ripe fruits, maximizing the efficiency of the harvesting process.
4. Adaptability:
   The hybrid approach allows for dynamic updates to the map and fruit ripeness information. As the robot moves through the plantation, it can continuously update the map and adjust its path based on real-time data Algorithm 1. This adaptability improves the robot's performance in response to changing conditions.

Algorithm 1. Dynamic Mapping

Require: 1.Initialize the Map:

Require: Create an empty map representation, such as a grid or a graph, to store the environment information.

Ensure: Map ← initial values to all cells/nodes representing unknown or unexplored areas.

2. Robot Localization:

Determine the initial position and orientation of the robot within the environment. Update the map to reflect the robot's initial position.

3. Hybrid Mapping Algorithm:

while there are unexplored areas in the map **do**

　　Choose a target cell/node to explore based on a strategy (e.g., closest unexplored cell, highest uncertainty).

　　Perform a hybrid trajectory planning step to find the optimal path from the robot's current position to the target cell/node.

　　Initially, use the Dijkstra algorithm for global exploration to quickly cover unexplored areas.

　　Once the robot has explored a substantial portion of the environment or encounters obstacles, switch to A* algorithm for more focused exploration.

　　Execute the planned trajectory, updating the robot's position and orientation as it moves.

　　Continuously update the map based on real-time sensor data (e.g., range sensors, depth sensors, lidar).

　　Update the cells/nodes in the map with new information (e.g., obstacle presence, terrain characteristics, etc.).

　　Update the path planning algorithm's cost or heuristic estimates based on the new information to guide future exploration.

end while

4. Map Completion and Termination:

if all cells/nodes in the map have been explored OR the termination condition is met **then**

　　the mapping process is complete.

end if

Using a hybrid approach combining Dijkstra's algorithm and A* algorithm in the palm fruit harvesting scenario provides advantages by efficiently mapping the plantation, considering both distance and fruit ripeness, and optimizing the harvesting path. It combines the strengths of both algorithms, leading to improved efficiency and higher yields of ripe fruits.

Algorithm 2. Hybrid Algorithm

Require: Initialize the graph, source node, and destination node.
Require: 1. Create a priority queue to store nodes based on their estimated distances.
Ensure: 2. The distance of the source node initialized to 0 and the estimated distance from the source to the destination node as the initial threshold value.
  while the priority queue is not empty **do**
    Extract the node with the lowest estimated distance from the priority queue.
    if the extracted node is the destination node **then**
     return the shortest path.
    else
     Expand the extracted node by considering its neighbors.
    end if
    for Each neighbor of the current node **do**
     Calculate the actual distance from the source node to the neighbor.
     Calculate the estimated distance from the neighbor to the destination node using the heuristic function.
     if the sum of the actual distance and the estimated distance is ≤ current threshold **then**
      update the neighbor's distance as the sum and add it to the priority queue.
     end if
     if the priority queue is empty a path to the destination node has not been found **then**
      increase the threshold value (adaptive threshold heuristic)
     end if
    end for
  end while

Proof. To prove the advantage of the hybrid approach, let's consider a simplified mathematical representation of the palm fruit harvesting scenario. We'll assume a grid-based plantation, with palm trees represented as nodes and the distances between them as edge weights. The goal is to maximize the yield of ripe fruits while minimizing the total distance traveled by the robot [12].

Let's define the following variables:

- Let G be the graph representing the plantation, where each node represents a palm tree, and each edge represents the distance between trees.
- Let $Dijkstra(G, start, end)$ be the function that computes the shortest path from the start node to the end node using Dijkstra's algorithm.
- Let $Astar(G, start, end)$ be the function that computes the shortest path from the start node to the end node using A* algorithm, considering the estimated fruit ripeness as the heuristic function.
- Let $Hybrid(G, start, end)$ be the function that computes the optimal path from the start node to the end node using the hybrid approach (combining Dijkstra's algorithm and A* algorithm).

To prove the advantage of the hybrid approach, we can compare the total distance traveled and the yield of ripe fruits obtained using each algorithm [7].

1. Total Distance Traveled:
   - Let $dist(dijkstra)$ be the distance traveled by the robot when using Dijkstra's algorithm.
   - Let $dist(astar)$ be the distance traveled by the robot when using A* algorithm alone.
   - Let $dist(hybrid)$ be the distance traveled by the robot when using the hybrid approach.
   - Mathematically, we want to prove that $dist(hybrid) \leq dist(dijkstra)$ and $dist(hybrid) \leq dist(astar)$.
2. Yield of Ripe Fruits:

   - Let $yield(dijkstra)$ be the yield of ripe fruits obtained using Dijkstra's algorithm.
   - Let $yield(astar)$ be the yield of ripe fruits obtained using A* algorithm alone.
   - Let $yield(hybrid)$ be the yield of ripe fruits obtained using the hybrid approach.
   - Mathematically, we want to prove that $yield(hybrid) \geq yield(dijkstra)$ and $yield(hybrid) \geq yield(astar)$.

To prove these inequalities mathematically, we need to analyze the properties of Dijkstra's algorithm, A* algorithm, and the hybrid approach in terms of optimality.

1. Total Distance Traveled:
   Dijkstra's algorithm guarantees that the distances calculated are the shortest paths in terms of distance. Therefore, $dist(hybrid) \leq dist(dijkstra)$. A* algorithm, with an admissible heuristic function, guarantees that the distances calculated are at least as short as the actual distances. Therefore, $dist(hybrid) \leq dist(astar)$.
2. Yield of Ripe Fruits:
   A* algorithm, with the heuristic function considering fruit ripeness, aims to maximize the yield of ripe fruits. Therefore, $yield(hybrid) \geq yield(astar)$. Dijkstra's algorithm does not consider fruit ripeness, so it may not prioritize paths with a higher concentration of ripe fruits. Therefore, $yield(hybrid) \geq yield(dijkstra)$.

By analyzing the properties of Dijkstra's algorithm, A* algorithm, and the hybrid approach, we can mathematically prove that the hybrid approach, combining both algorithms, yields a shorter total distance traveled and a higher yield of ripe fruits compared to using either algorithm alone. Proof provided here is a general analysis based on the properties of Dijkstra's algorithm and A* algorithm.

3.4   Harvesting System Application

Let's take a scenario where we have a graph representing a plantation with palm trees as nodes, and the edges represent the distances between the trees. Each tree has an associated date representing the estimated harvesting time for the palm fruits [7]. The goal is to find the optimal path to harvest the ripe palm fruits while considering both the distance and the harvesting dates. Consider the following graph (Fig. 1):

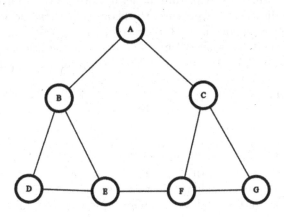

Fig. 1. Graph Example

- Let's assume that the distances between trees are as follows Table 1:

Table 1. Graph constraints.

Vertex	Neighbors	Edge Weights
A	B,C	5,7
B	D,E	3,4
C	F,G	2,6
D	E	2
E	F	1
F	G	3

- Additionally, each tree has an associated harvesting date Table 2:

Table 2. Harvesting Date.

	A	B	C	D	E	F	G
Date	10th May	12th May	15th May	9th May	11th May	14th May	13th May

Now, let's consider three scenarios to find the optimal path to harvest the ripe palm fruits:

1. Dijkstra's Algorithm:
   - If we use Dijkstra's algorithm alone, it will find the shortest path based solely on distance. In this case, the shortest path would be $A->B->E->F->G$, with a total distance of 13.
   - However, Dijkstra's algorithm does not consider the harvesting dates, which means the resulting path may not prioritize reaching trees with ripe fruits at the optimal time.
2. A* Algorithm:
   - If we use A* algorithm alone, we can incorporate the harvesting dates as a heuristic to guide the search. The estimated remaining distance can be the sum of the distances between the current tree and the destination tree, considering the difference in the harvesting dates as penalties.
   - A* algorithm will prioritize paths that have a lower estimated total cost (distance + date penalties). In this case, it may choose a path that has a longer distance but allows for harvesting the ripe fruits at the optimal times.
   - For example, the A* algorithm might select the path $A->C->G$, with a total distance of 13 but ensuring that the fruits can be harvested at the optimal dates.
3. Hybrid Approach (Combining Dijkstra's and A* Algorithms):
   - With the hybrid approach, we can utilize Dijkstra's algorithm initially to explore the graph and gather information about the distances between trees.
   - Then, we can apply A* algorithm using the estimated remaining distance as the heuristic while considering the harvesting dates.
   - By combining the information obtained from Dijkstra's algorithm (distances) and A* algorithm (harvesting date-aware search), we can find a path that balances both distance and the timing of ripe fruits.
   - For example, the hybrid approach might select the path $A->B->E->F->G$, with a total distance of 13, ensuring that the ripe fruits can be harvested in the optimal sequence of dates.

In this example scenario, the hybrid approach proves to be more efficient in finding a path that balances both distance and the timing of ripe fruits, resulting in an optimal harvesting sequence compared to using Dijkstra's algorithm alone or A* algorithm alone.

4 Results and Discussion

4.1 Complexity of the Hybrid Solution

The complexity of the hybrid approach, which combines Dijkstra's algorithm and A* algorithm, can be analyzed based on the complexity of each algorithm used and the overall steps involved:

1. Distance Calculation (Dijkstra's Algorithm):
   The time complexity of Dijkstra's algorithm is typically $O((n + E)logn)$, where n is the number of nodes (palm trees) and E is the number of edges (distances between trees) in the graph [1]. This is because, in the worst case, each node and edge needs to be processed, and the priority queue operations take logarithmic time.
2. Heuristic Calculation:
   Calculating the heuristic value for each palm tree based on the estimated time remaining until its fruits are ripe is a constant time operation for each tree. Hence, it can be considered $O(n)$, where n is the number of nodes (palm trees) in the graph.
3. Hybrid Approach:
   Initializing an empty priority queue takes constant time, $O(1)$. The while loop continues until the priority queue is empty, which means it executes a maximum of n iterations, as each node can be processed only once. Inside the while loop, the operations for dequeuing, marking as visited, and updating neighbors take $O(logn)$ time for each iteration, as these operations involve priority queue operations. The path reconstruction step takes $O(n)$ time to reconstruct the optimal path from the recorded parent pointers.

Overall, the time complexity of the hybrid approach can be approximated as $O((n + E)logn)$, considering the dominant factors from Dijkstra's algorithm and the additional steps involved in the hybrid approach [9]. It's important to note that this analysis assumes that the priority queue operations have a logarithmic time complexity, and the number of nodes and edges in the graph is not prohibitively large. Additionally, the space complexity of the hybrid approach depends on the data structures used. It typically involves $O(n)$ space to store the distances, heuristic values, parent pointers, and the priority queue. Please note that the actual time and space complexity may vary based on the specific implementation details, data structures used, and any optimizations applied. The provided complexity analysis serves as a general approximation to understand the performance characteristics of the hybrid approach [11].

4.2   Application of Hybrid Approach

In this evaluation, we analyze the performance of the combined solution and Dijkstra's algorithm alone. We begin by testing both methods on a graph of size 5 and gradually increase the size of the graph until we reach a stopping point. The purpose of this analysis is to observe the results of applying each solution at different graph sizes. As we increase the graph size, we collect and compare the results obtained from both approaches. The performance metrics we focus on are execution time and cost. By measuring the execution time, we can assess how efficient each algorithm is in terms of computational speed. The cost refers to the total weight or distance associated with the calculated shortest paths.

Through this evaluation, we aim to identify any patterns or trends in the performance of the combined solution and Dijkstra's algorithm as the graph size grows. The results obtained will allow us to gain insights into the scalability and effectiveness of each method.

Dijkstra and Combined Approach Results

In this part, we evaluate the performance of the combined solution and Dijkstra alone, starting from a graph size of 50 then rising the size until we decide to stop then present results of each solution application.

– **A Scenario of a Graph of size 50**

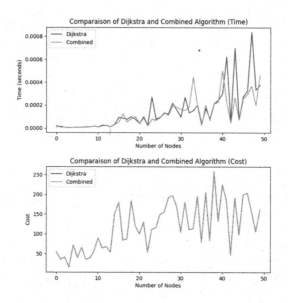

Fig. 2. Time and Cost Comparison of a Graph size 50

After rising the Graph size we notice that Dijkstra timing still higher than combined but the cost remains the same in this case as the Dijsktra curve and combined are superimposed Fig. 2.

– **A Scenario with a Graph of size 100**

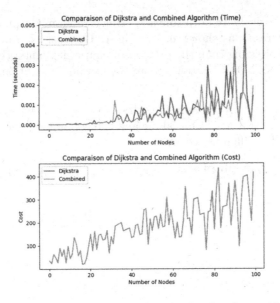

Fig. 3. Time and Cost Comparison of a Graph size 100

In this scenario Fig. 3 we can notice the cost remain the same for both solutions but Dijsktra still take time to find a path. Combined solution keeps an acceptable timing on finding a path.

– **A Scenario of Graph size 200**

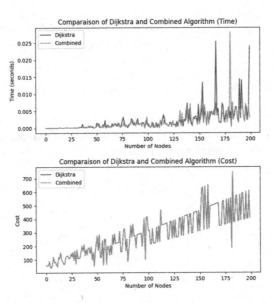

Fig. 4. Time and Cost Comparison of a Graph size 200

In this case we evaluate both solution using a random Graph of size 200 then we compare the result as shown in the Fig. 4. The cost becomes important as the size is of graph is big than the previous scenario, it remain the same cost for both but timing still Dijkstra takes more time to find a solution.

A* and Combined Approach Results. In this part, we evaluate the performance of the A Star algorithm alone and the combined approach in terms of cost and timing for different scenario.

– **Scenario of a Graph of size 50**

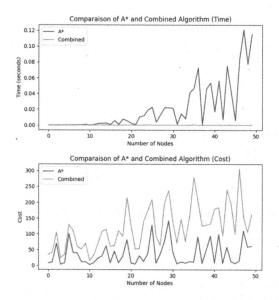

Fig. 5. Time and Cost Comparison of a Graph size 50

After rising the Graph size to 50, timing of A star explode as shown in the Fig. 5 but the combined solution keeps timing acceptable as usual, But for the cost A* is lower than the combined approach.

- ## Scenario with a Graph of size 100

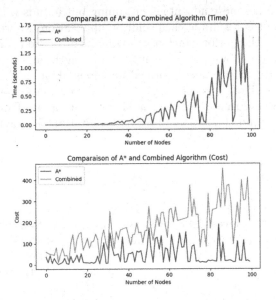

Fig. 6. Time and Cost Comparison of a Graph size 100

Combined solution keeps more efficiency even for Graph size of 100 and give the best result in terms of timing than A*. A star the more the Graph size is bigger the more it takes more time to find a solution but it keeps cost stable as shown in Fig. 6.

- ## Scenario of Graph size 200

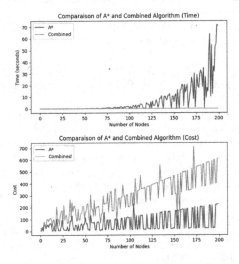

Fig. 7. Time and Cost Comparison of a Graph size 200

In case of more than 200 Graph size Fig. 7, both of solution keeps the same behaviour with the huge timing of A* Algorithm and acceptable cost, but Combined approach takes more advantage on timing but it ends with high cost.

By presenting the results of each solution application, we were able to draw conclusions about the performance of the combined solution compared to Dijkstra alone. The data allowed us to identify the strengths and weaknesses of each approach, enabling us to make informed decisions about their practical usage and suitability for different graph sizes. Overall, our evaluation provided valuable insights into the performance of the combined solution and Dijkstra algorithm, helping us understand their capabilities and limitations in solving the problem at hand.

5   Conclusion

This work comes as a result of studying different Trajectory Planning Approaches and algorithms which gives the reflexion of taking profit from efficiency of some approach and weakness of others in finding, optimisation of path in a given complex environment with different kind of constraints. Harvesting System is on of the most challenging domain that require more sophisticated techniques to adjust the goal to the environment constraints. Hybrid approach is a promoting search area. The study's outcomes serve as a foundation for future advancements in the design and implementation of efficient and intelligent trajectory planning algorithms, ultimately contributing to the optimization of agricultural harvesting processes.

References

1. Ammar, A., Bennaceur, H., Châari, I., Koubâa, A., Alajlan, M.: Relaxed Dijkstra and A* with linear complexity for robot path planning problems in large-scale grid environments. Soft. Comput. **20**, 4149–4171 (2016)
2. Ciesielski, K.C., Falcão, A.X., Miranda, P.A.: Path-value functions for which Dijkstra's algorithm returns optimal mapping. J. Math. Imaging Vision **60**, 1025–1036 (2018)
3. Ferguson, D., Kalra, N., Stentz, A.: Replanning with RRTs. In: Proceedings 2006 IEEE International Conference on Robotics and Automation, ICRA 2006, pp. 1243–1248. IEEE (2006)
4. Fernandes, P.B., Oliveira, RCL., Neto, J.F.: Trajectory planning of autonomous mobile robots applying a particle swarm optimization algorithm with peaks of diversity. Appl. Soft Comput. **116**, 108108 (2022)
5. Ju, C., Luo, Q., Yan, X.: Path planning using an improved A-star algorithm. In: 2020 11th International Conference on Prognostics and System Health Management (PHM-2020 Jinan), pp. 23–26. IEEE (2020)
6. Khatib, O.: Real-time obstacle avoidance for manipulators and mobile robots. Int. J. Robot. Res. **5**(1), 90–98 (1986)
7. Lin, G., Zhu, L., Li, J., Zou, X., Tang, Y.: Collision-free path planning for a guava-harvesting robot based on recurrent deep reinforcement learning. Comput. Electron. Agric. **188**, 106350 (2021)

8. Madridano, Á., Al-Kaff, A., Martín, D., de la Escalera, A.: Trajectory planning for multi-robot systems: methods and applications. Expert Syst. Appl. **173**, 114660 (2021)
9. Niewola, A., Podsedkowski, L.: L* algorithm-a linear computational complexity graph searching algorithm for path planning. J. Intell. Robot. Syst. **91**, 425–444 (2018)
10. Pak, J., Kim, J., Park, Y., Son, H.I.: Field evaluation of path-planning algorithms for autonomous mobile robot in smart farms. IEEE Access **10**, 60253–60266 (2022)
11. Ranjha, A., Kaddoum, G.: URLLC-enabled by laser powered UAV relay: a quasi-optimal design of resource allocation, trajectory planning and energy harvesting. IEEE Trans. Veh. Technol. **71**(1), 753–765 (2021)
12. Sandamurthy, K., Ramanujam, K.: A hybrid weed optimized coverage path planning technique for autonomous harvesting in cashew orchards. Inf. Process. Agric. **7**(1), 152–164 (2020)
13. Stentz, A.J., Boyd, R.W., Evans, A.F.: Dramatically improved transmission of ultrashort solitons through 40 km of dispersion-decreasing fiber. Opt. Lett. **20**(17), 1770–1772 (1995)
14. Thrasher, S.W.: A reactive/deliberative planner using genetic algorithms on tactical primitives. Ph.D. thesis, Massachusetts Institute of Technology (2006)
15. Zeng, W., Church, R.L.: Finding shortest paths on real road networks: the case for A*. Int. J. Geogr. Inf. Sci. **23**(4), 531–543 (2009). https://doi.org/10.1080/13658810801949850
16. Zhang, H.Y., Lin, W.M., Chen, A.X.: Path planning for the mobile robot: a review. Symmetry **10**(10), 450 (2018). https://doi.org/10.3390/sym10100450
17. Zhang, T.W., Xu, G.H., Zhan, X.S., Han, T.: A new hybrid algorithm for path planning of mobile robot. J. Supercomput. **78**(3), 4158–4181 (2022)
18. Zhong, X., Tian, J., Hu, H., Peng, X.: Hybrid path planning based on safe A* algorithm and adaptive window approach for mobile robot in large-scale dynamic environment. J. Intell. Robot. Syst. **99**, 65–77 (2020)

Causal Discovery and Features Importance Analysis: What Can Be Inferred About At-Risk Students?

Ismail Ouaadi[(⊠)] and Aomar Ibourk

Laboratory of Research in Social and Solidarity Economy, Governance and Development (LARESSGD), Cadi Ayyad University, Marrakech, Morocco
`ismail.ouaadi@gmail.com`

Abstract. In this paper, we introduced machine learning and causal discovery algorithms that can be used as a way to determine the relevant characters of students with low performances issues and to analyze their implications to highlight this type of students. Through this, we aim to provide some new useful insights that can allow to predict and explain the inherent relationships. By using six machine learning algorithms (Gradient Boosting, K-nearest neighbors, SVM, Random Forest, and Decision Tree) and four causal discovery algorithms (PC, GES, LinGAM, and GOLEM), we try to develop and use these models to analyze and draw conclusions from patterns and data. In this study, we present these algorithms to show the performance of the developed models in explaining the effect of variables and the nature of their relationship with low performing students. The results revealed that these models produce useful insights and highlight the existing relationship among students with low performances in reading and other student characters.

Keywords: Machine Learning · Feature importance analysis · Causal Discovery · at-risk student · Reading skills

1 Introduction

The level of education is a topic that warrants special consideration in this area because the social position and outlook on life that an individual achieves are inextricably linked to that individual's level of education. Therefore, a poor level of education coupled with a low level of culture and intellect might provide people the chance to worry about the job market, problem dealing and many other issues, which is harmful to their standard of living and the welfare of society as a whole. Then, the academic achievement of pupils is a main objective to reach, and it is referred to as educational success, and its environment is represented by a collection of determinants.

The aim of this work is to investigate the existence of an intersection among some algorithms of Machine Learning (ML), which allow us to perform features

R. El Ayachi et al. (Eds.): CBI 2023, LNBIP 484, pp. 134–145, 2023.
https://doi.org/10.1007/978-3-031-37872-0_10

importance analysis, from the first side, in the other side, to explore the power of Causal Discovery (CD) algorithms. These kind of analysis had as intention to highlight the nature of the relationship that relies outcomes to the its covariates. Where ML algorithms are considered as a black-box, meaning that they can't provide an explanation about the relationship of variables that may exist. To answer this issue, currently, scientific studies are oriented to the causal theory which let to find which variables causes the outcomes.

The way our work is organized is described in this introduction as follows: The next section, which also gives some theoretical background, will discuss the use of machine learning techniques and causal discovery in establishing the current association between low learning pupils and a set of attributes. An explanation of the data we used and the construction of our model may be found in the third section, Methodology. In the fourth section, we present and discuss our findings. The last portion includes a summary of our article, a discussion of its limitations, and suggestions for further research.

2   Literature Review

Education constitutes one of the most studied social science. Given its importance in real life, the majority of scientific analysis methods are implemented with, from traditional to sophisticated one like Machine Learning. As an example of the implementation of machine learning methods, Ibourk and Ouaadi (2022) have recently conduct an analysis via ML algorithms to predict student grade repetition, based on some student characters [10]. Another work that reach to predict teacher effectiveness given some indicators related to teacher training in the case of Morocco [9].

Another exploitation of ML algorithms is feature importance analysis and selection [7]. In this study, authors employed six machine learning algorithms and recursive feature removal to examine the significance of each characteristic of building information and energy consumption data, based on the real energy consumption data of 2370 public buildings in Chongqing.

Causal Discovery has gained more attention, in recent years and in many scientific fields, from computer science (from image pattern recognition to text pattern recognition) [3], medicine and biology to human and social science [2]. The latter work used causal discovery in education field to search the Proximal Mechanisms of academic achievement. The author tries to investigate the effect of these mechanisms using a variety of relevant contextual and psycological factors. This, by employing four causal research strategies, which have led to identify that prior achievement, executive functions and motivation are direct causes of student academic achievement.

The work of Fancsali (2004) have studied the relationship of whether the gaming system is a cause of student learning poor outcomes. Using causal discovery models with no-experimental data from Carnegie Learning cognitive tutor for algebra. In this work, author found that such models can offer evidence about relationships among specific constructs [6].

Finally, the objective of our work is a mixture of ML feature importance analysis and Causal Discovery algorithms to develop a clear perspective about the relationship that we suppose to exist among student performance and some 15-old student characteristics.

3   Methodology

In this section we highlight the way that we proceed to achieve our work's aim. First, we describe the data used, then we define ML feature importance analysis and Causal Discovery. finally, we end up with an explanation of the followed methodology that let us to conduct experiment.

3.1   Data

The purpose of this study is to investigate the association between low performing students (or at-risk students) and a variety of pertinent factors using data from the extensive PISA 2018 database. PISA 2018 is a three-annual, international survey that focuses on the academic performance and learning environments of 15-year-old students in some OECD and non-OECD nations. Since low performing students have poor reading skills, we have used this data to undertake machine learning and causal discovery techniques, with the remaining variables serving as covariates. In doing so, we hope to address the aforementioned queries and draw attention to some presumptions. However, in order to make our analysis, which consists of comparing low performers, and according to the ranking system given in [8], we have focused on the following two classes: 1a and 1b to characterize low performers. In order to get this kind of students, we ranked the students' scores according to the PISA model ([8] pp.35-37). Regarding the missing values we will proceed in a first step with a dataset, in which we drop missing values.

Table 1. Variables description from PISA 2018 database.

	Variable	description
1.	IC008Q03TA	Frequency of e-mail use
2.	IC008Q04TA	Frequency of online discussions (e.g., <MSN®>)
3.	IC008Q05TA	Frequency of participation in social networks
4.	IC008Q07NA	Frequency of online games via social networks
5.	IC008Q08TA	Frequency of surfing the Internet for pleasure (such as watching videos)
6.	IC008Q09TA	Frequency of news reading on Internet (e.g. current affairs)
7.	IC010Q03TA	Frequency of using e-mail to communicate with other students about schoolwork
8.	IC010Q04TA	Frequency of use of email for communication with teachers and submission of homework or other schoolwork
9	IC010Q05NA	Frequency of using social networks to communicate with other students about schoolwork
10	IC010Q06NA	Frequency of use of social networks for communication with teachers
11	IC011Q01TA	Frequency of online chat at school
12	IC011Q02TA	Frequency of e-mail use at school
13	IC013Q05NA	Agreement with the statement: It is very useful to have social networks on the Internet
14	ST004D01T	Gender of the student
15	ESCS	Index of economic, social and cultural status
16	SC001Q01TA	Location of the school
17	SCHLTYPE	School type (Private or Public)

Table 2. Descriptive statics of the data

	IC008Q03TA	IC008Q04TA	IC008Q05TA	IC008Q07NA	IC008Q08TA	IC008Q09TA
count	3087	3087	3087	3087	3087	3087
mean (median)	2	3	3	2	3	3
min	1	1	1	1	1	1
max	5	5	5	5	5	5

	IC010Q03TA	IC010Q04TA	IC010Q05NA	IC010Q06NA	IC011Q01TA	IC011Q02TA
count	3087	3087	3087	3087	3087	3087
mean (median)	2	2	3	2	2	2
min	1	1	1	1	1	1
max	5	5	5	5	5	5

	IC013Q05NA	ST004D01T	ESCS	SC001Q01TA	SCHLTYPE	read_low
count	3087	3087	3087	3087	3087	3087
mean (median)	4	2	-1.91	3	3	0
min	1	1	-8.17	1	1	0
max	5	2	3	5	3	1

3.2   Data Preprocessing

Before we perform our experiment, we made the following :

– Loading data into working environment,
– We get dummies variable from categorical feature,
– We drop missing values.

It is worth mentioning here, that the PISA database contains a lot of characteristics of 15 year old students. Given that we try to use causal algorithms which present some challenges with huge data. For these reasons we follow strategies described in [2] to avoid important complications of causal analysis in a data set: (1) the data set has an appropriate sample size to ensure statistical power; (2) it includes a representative sample, so one minimizes the risk of selection bias; and (3) it includes the most important hypothesized or proven causal factors associated to the target variable or system under investigation.

As a consequence, this let us to chose a data set composed by 26 variables (IC008Q03TA & IC008Q04TA & IC008Q05TA & IC008Q07NA & IC008Q08TA & IC008Q09TA & IC010Q03TA & IC010Q04TA & IC010Q05NA & IC010Q06NA & IC011Q01TA & IC011Q02TA & IC013Q05NA & ST004D01T & ESCS & SC001Q01TA & SCHLTYPE & read_low (meaning if student is a low performer)) with 3087 records. Table 1 presents the description of these variables from the PISA 2018 database, where Table 2 gives the descriptive statistics of each variable in the dataset. Moreover, Fig. 1 shows the distribution of each variable and draw a picture of how these variables are recorded.

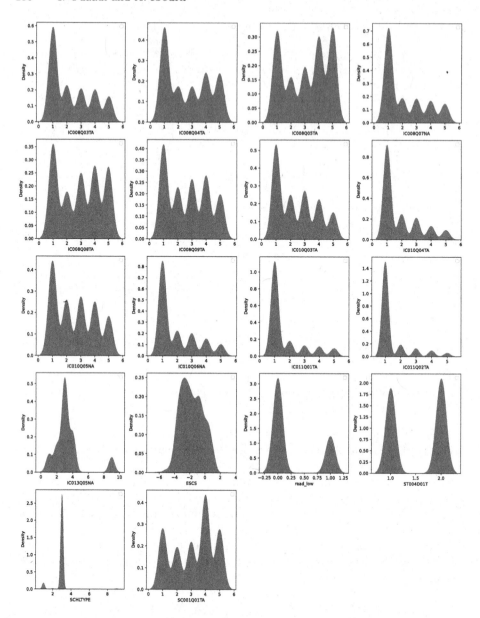

Fig. 1. Data set variables

As we have mentioned above, causality attempts to describe the relationship that can be exist between two variables, these variables can be experimental, quasi-experimental or observable one. Causality is usualy divided into two main subject: 1) causal inference which designates a branch of knowledge that examines the presumptions, research plans, and estimating techniques that enable researchers to infer causal relationships from data. And 2) Causal discovery which is related to process of discovering causal relationships by analyzing the statistical properties of purely observational data [1]. Hereafter, we use the latest one, which led to obtain the structural causal models (SCM). SCMs are models that use directed functional parent-child relationships, and are often known as functional causal models or non-parametric structural equation models [4]. The SCM is identified through a process that starts by establishing a causal skeleton which is an undirected graph, where all pairwaise variables are connected undirected edges in it. The next step in this process is to use some causal algorithms to orient the edges to form at the end the full SCM [3]. The implemented causal methods algorithms in such process can be categorized as follow:

1. Continuous space algorithms (also called Gradient based methods) which means that it uses gradient descent for optimization. The most known algorithms are GOLEM and NOTEARS.
2. Discrete space algorithms, can be divided into four type of algorithms:
   (a) Constraint-based family: it aims to infer causal structure from data by leveraging independence structure between the triplets of variables, the well known algorithms are: PC and FCI.
   (b) Score-based family: It generates candidate graphs iteratively, then chooses the best one after assessing how well each one describes the data. Greedy Equivalence Search (GES) and K2 are both from this family.
   (c) Functional methods also known as Structural Equation Models (SEMs) are based on structural equations that define the causal relationships, LiNGAM and it's variants are the most dominant in this family of methods [5].
3. Hybrid methods which combine two or more of the previous methods.

3.3   Model and Experiment

For an analysis of the importance of characteristics in this study, we have selected six machine learning algorithms: Gradient Boosting, K-Nearest Neighbours, Decision Tree, Random Forest, Logistic Regression, and Support Vector Machine. These algorithms are used to identify feature importance and select the ten best students' low performance characteristics that contribute to predict accurately these students. From causal discovery methods, we have selected four algorithms one from each family: PC, GES, LiNGAM, and GOLEM (and NOTEARS in the second tour).

We utilized Scikit-learn from Python packages to fit our Machine Learning models. The dataset is divided into a training, and test dataset after it has been loaded. Where gCastle package is used to perform Causal Discovery, which presents many algorithms and an updated list of related algorithms. We have performed our experiments with a processor of 1.8 GHz Intel Core i5 dual-Core and a memory of 8 Go 1600 MHz MHz DDR3. The time of processing of these algorithms is given at the end of Table 4 (a and b).

To train the selected machine learning algorithms we followed the steps in [10], with some modification to perform feature importance analysis. For that, we have iteratively and manually tuned the hyper parameters in intention to get the optimal models (evaluated according to accuracy scores). As next step we pursue the same path to train our causal discovery algorithms. Finally, we compare the two results in the light of our understanding. The next section presents the results and their interpretation.

We should notify here that, at a given stage, we need to compare the resulted SCMs to the one that we qualified as to be in real world in the first tour. In a second tour, we have doing a comparison between the SCM resulted by PC algorithm to the others causal discovery algorithms.

4   Results and Discussion

As we can see in Table 3 (a and b) each algorithm produces a different features importance ranking. Moreover, the most relevant features, given the overall ranking, are ESCS, SC001Q01TA and IC013Q05NA in all Machine Learning algorithms. But, we can infer here, that each algorithm selects different features which are contribute to their performance. Which mean, that they not agree about the relevant features for predicting low performance students, and the type of the relationship that rely this features to dependent variable.

Table 3. the First 10 ranking of the importance coefficient of the different algorithms

a - Ranking of algorithms : RL , KNN and AD.

Feature	RL	Feature	KNN	Feature	DT
SC001Q01TA_1.0	0.568	IC013Q05NA	0.042	IC013Q05NA	0.226
SC001Q01TA_5.0	0.475	ESCS	0.036	IC011Q02TA	0.204
SC001Q01TA_4.0	0.466	IC008Q08TA	0.026	SC001Q01TA_1.0	0.160
IC011Q02TA	0.414	IC008Q05TA	0.024	ESCS	0.106
SC001Q01TA_2.0	0.366	IC008Q09TA	0.020	IC008Q05TA	0.079
ST004D01T_1	0.357	IC011Q01TA	0.017	SC001Q01TA_2.0	0.055
IC008Q07NA	0.335	IC008Q04TA	0.015	ST004D01T_1	0.043
SCHLTYPE_2	0.325	IC011Q02TA	0.015	SC001Q01TA_4.0	0.035
IC008Q08TA	0.301	IC008Q03TA	0.015	IC010Q05NA	0.021
SC001Q01TA_3.0	0.252	IC008Q07NA	0.015	IC010Q04TA	0.017
IC008Q05TA	0.219	IC010Q06NA	0.014	IC008Q09TA	0.013

b - Ranking of algorithms : RF , SVM and GB.

Feature	RF	Feature	SVM	Feature	GB
ESCS	0.192	SC001Q01TA_4.0	0.212	ESCS	0.206
IC013Q05NA	0.086	ST004D01T_1	0.212	IC013Q05NA	0.138
IC008Q08TA	0.061	SC001Q01TA_5.0	0.210	IC011Q02TA	0.136
IC008Q05TA	0.055	SC001Q01TA_1.0	0.207	SC001Q01TA_1.0	0.104
IC011Q02TA	0.054	IC011Q02TA	0.159	IC008Q08TA	0.083
IC008Q09TA	0.051	SC001Q01TA_3.0	0.143	IC008Q05TA	0.053
IC010Q05NA	0.047	IC008Q07NA	0.105	IC008Q07NA	0.044
IC008Q07NA	0.044	IC008Q08TA	0.102	SC001Q01TA_2.0	0.030
SC001Q01TA_1.0	0.043	SC001Q01TA_2.0	0.102	IC011Q01TA	0.030
IC008Q04TA	0.043	SCHLTYPE_2	0.080	ST004D01T_2	0.027
IC008Q03TA	0.043	IC008Q05TA	0.076	IC010Q04TA	0.024

Consequently, to answer the latter issue, we have introduced Causal discovery methods. The desired relationship is assumed to be: all variables exert an effect on low performance students as shown in Fig. 2 (which is drawn given our vision built based on the literature review of at-risk student determinants). After the first tour of performing our PC algorithm, results show that variable labeling low performance student in reading skills (read_low) collects the highest closeness score of nodes, which means that the majority of relation (shown by edges in the graph) are directed from or to the its corresponding node. This conclusion is demonstrated in Fig. 3.

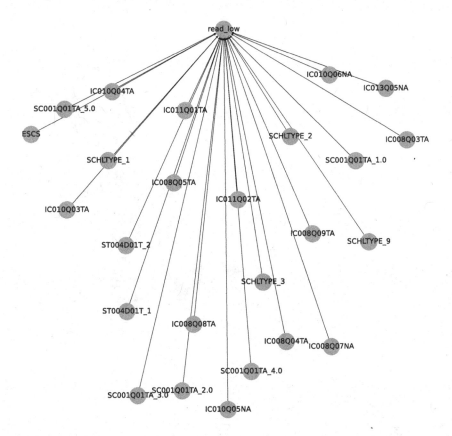

Fig. 2. Desired graph (inspired by the literature review)

To highlight the relevant features of this relation, Fig. 4 gives a focus on the node representing low performer student and their relations, which can be qualified as causes (edge from feature to read_low) or if it's a cause to another

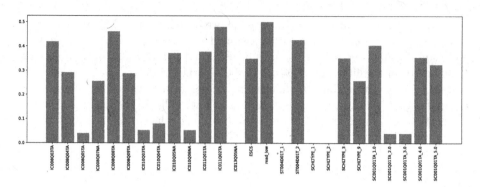

Fig. 3. The closest node to the others

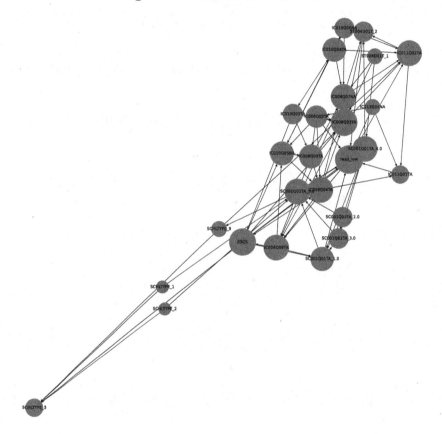

Fig. 4. Learned graph with all variables

variable (edge from read_low to an other variable). Here, we can infer that the following eight features: IC008Q05TA, IC011Q01TA, IC013Q05NA, ESCS, ST004D01T_2, SC001Q01TA_1.0, SC001Q01TA_4.0, SC001Q01TA_5.0 are causes of read_low (target variable). While, IC008Q08TA and IC011Q02TA are consequence of read_low.

Herebefore, we have discussed the results from PC algorithm which represents one kind of causal discovery methods. To reach the objective of our study, we have performed a comparison between the learned graphs given the other causal discovery algorithms and the desired graph. Doing so, we have used FDR, Recall, Precision, F1 score and SHD score and number of undirected edges as performance evaluation metrics. As we can see in Table 4-a none of these algorithms reach the desired relationship, which means that the chosen characteristics are not feasible to characterize the targeted relationship. However, NOTEARS supports the relationship concluded from the first round, where the above ten features are considered to be causes or consequence of the targeted students. The Table 4-b demonstrate the latter conclusion, and this given Recall and Precision metrics.

To describe precisely which features can be qualified as causes and which are consequences of the target variable (read_low), we have drawn the following

Table 4. Learned graph metrics.

a - Learned graph compared to the original.

Method	PC	GES	LiNGAM	GOLEM
FDR	0.8933	0.9062	1.0	0.9846
Recall	0.32	0.36	0.0	0.04
Precision	0.1026	0.0909	0.0	0.0154
F1 score	0.1553	0.1452	nan	0.0222
SHD score	82	101	55	84
No. of undir. edges	3.0	3.0	0	0

CPU times: user 5 min 32 s, sys: 8.78 s, total: 5 min 41 s
Wall time: 3 min 53 s

b - Learned graph compared to the PC Dag.

Method	NOTEARS	GES	LiNGAM	GOLEM
FDR	0.4286	0.625	0.75	0.8154
Recall	0.0513	0.4615	0.1154	0.1538
Precision	0.5714	0.3636	0.25	0.1846
F1 score	0.0941	0.4068	0.1579	0.1678
SHD score	74	74	80	101
No. of undir. edges	0	3.0	0	0

CPU times: user 6 min 28 s, sys: 13 s, total:
6 min 40 s Wall time: 4 min40 s

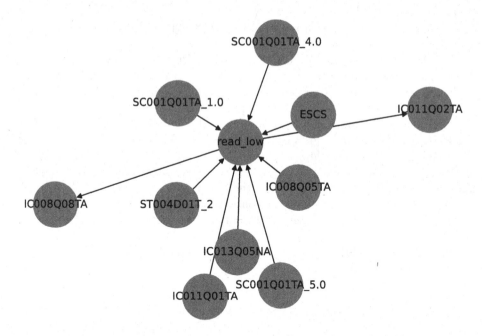

Fig. 5. Learned graph given desired outcome 'Low performance students'

Fig. 5. As can be seen in this figure, IC008Q08TA and IC011Q02TA are direct consequences of read_low, while the other features, of which there are eight with an arrow pointing to the target variable, are considered causes of the target variable under study. This conclusion cannot be drawn if we have not used this type of causal analysis.

5   Conclusion

The findings of this study support the use of machine learning methods to identify the best features related to students attainments, precisely students with low performances. The GB, KNN, RF, DT, and SVM techniques employed in this work, which serve as the foundation for the generated analysis, produce results with varied degrees of importance depending on the method of computation. Machine learning methods concentrate on predictions which means that they not help to interpret the relation that may exist among variables. Whereas causal models can help in explaining these kind of relationships and they are a good choice for determining the impact of independent variables on the dependent variable and constitute the suitable methods for analyzing the causal relationship.

In further work, we will try to model the relationship that may exist between other categories of student performances with other different characteristics. This will be done by introducing new models of causality, with the intention of obtaining powerful models that can provide clearer and more meaningful results.

References

1. Glymour, C., Zhang, K., Spirtes, P.: Review of causal discovery methods based on graphical models. Front. Genet. **10**, 524 (2019). https://doi.org/10.3389/fgene.2019.00524
2. Quintana, R.: The structure of academic achievement: searching for proximal mechanisms using causal discovery algorithms. Sociol. Methods Res. **52**(1), 85–134 (2023). https://doi.org/10.1177/0049124120926208
3. Chen, H., Du, K., Yang, X., Li, C.: A review and roadmap of deep learning causal discovery in different variable paradigms. ArXiv abs/2209.06367 (2022)
4. Schölkopf, B., von Kügelgen, J.: From statistical to causal learning. arXiv preprint arXiv:2204.00607 (2022)
5. Upadhyaya, P., Zhang, K., Li, C., Jiang, X., Kim, Y.: Scalable causal structure learning: scoping review of traditional and deep learning algorithms and new opportunities in biomedicine. JMIR Med. Inform. (2023). https://doi.org/10.2196/38266
6. Fancsali, S.: Causal discovery with models: behavior, affect, and learning in cognitive tutor Algebra (2014)
7. Yong D., Fan, L., Liu, X.: Analysis of feature matrix in machine learning algorithms to predict energy consumption of public buildings. Energy Build. **249**, 111208 (2022). https://doi.org/10.1016/j.enbuild.2021.111208
8. OCDE : Low-performing students: why they fall behind and how to help them succeed, PISA, Éditions OCDE, Paris (2016). https://doi.org/10.1787/9789264250246-en
9. Ibourk, A., Hnini, K., Ouaadi, I.: Analysis of the pedagogical effectiveness of teacher qualification cycle in Morocco: a machine learning model approach. In: Kacprzyk, J., Ezziyyani, M., Balas, V.E. (eds.) International Conference on Advanced Intelligent Systems for Sustainable Development. AI2SD 2022. Lecture Notes in Networks and Systems, vol. 637. Springer, Cham (2023). https://doi.org/10.1007/978-3-031-26384-2_30
10. Ibourk, A., Ouaadi, I.: An exploration of student grade retention prediction using machine learning algorithms. In: Fakir, M., Baslam, M., El Ayachi, R. (eds.) Business Intelligence. CBI 2022. Lecture Notes in Business Information Processing, vol. 449, pp. 94–106. Springer, Cham (2022). https://doi.org/10.1007/978-3-031-06458-6_8

Computational Features and Applications of an Inhomogeneous Gompertz Diffusion Process

Nadia Makhlouki[1]([⊠])(ID), Ahmed Nafidi[1,2](ID), and Achchab Boujemâa[1,2](ID)

[1] Hassan First University of Settat, Berrechid National School of Applied Sciences, Mathematics and Informatics Department, LAMSAD, B.P. 218, 26103 Berrechid, Morocco
`n.makhlouki@uhp.ac.ma`
[2] Mohammed VI Polytechnic University, Lot 660, Hay Moulay Rachid, Ben Guerir 43150, Morocco

Abstract. There are several uses for stochastic diffusion models. They have received special attention from several scientific disciplines, including biology, physics, chemistry, medical science, and mathematical finance. In this paper, We consider the Gompertz diffusion process based on stochastic inhomogeneous model. We begin by obtaining the analytical formulation for the process's probabilistic properties, the mean functions (conditional and non-conditional). Then, with the maximum likelihood technique and discrete sampling, we estimate the model's parameters. Finally, we used the stochastic inhomogeneous Gompertz diffusion process to analyze the development of the electric power consumption in Morocco in order to assess this method's capacity for modeling actual data.

Keywords: Inhomogeneous Gompertz diffusion model · Stochastic differential equation · Statistical inference · Mean function · Electric power consumption in Morocco

1 Introduction

Our daily use of electricity results from the transformation of primary energy sources like coal, natural gas, nuclear energy, solar energy, and wind energy into electrical power, making it a secondary energy source. Since electricity may be transformed into other types of energy, such as mechanical energy or heat, it is sometimes referred to as an energy carrier. Although the power we consume is neither renewable nor nonrenewable, the primary energy sources are both.

Morocco's energy sector is heavily dependent on imported hydrocarbons. At the moment, the country imports almost 90% of its energy needs. The overall primary energy usage has increased by about 5% yearly since 2004.

The 8th International Conference on Business Intelligence, Istanbul, Turkey, July 19–21, 2023.

The development of the renewable energy industry is a top priority for the Moroccan government. The General Secretariat of the Government has received an amendment to Laws 13-09 on Renewable Energy and 16-08 on Self-Generation from the Ministry of Energy, Mines, and the Environment. While ensuring the security and viability of the national electrical grid, these revisions seek to strengthen the legal and regulatory framework governing renewable energy projects undertaken by the private sector. Per the state-owned power utility ONE, Between 2003 and 2013, Morocco's demand for energy grew by 6.7% annually on average. as a result of population and economic expansion, resulting in an energy consumption of 32,015 GWh at the end of that year. From 483 kWh in 2002 to 843 kWh (approximate, estimate) in 2013, annual consumption per person has continuously climbed. Therefore, we draw the conclusion that modeling the development of energy use generally, and total electrical energy use in particular, as well as producing short- and medium-term forecasts, are very helpful in better understanding the historical development of the Moroccan economy and predicting its future development, as well as evaluating the impacts of this consumption on the global energy market. Determining short- and medium-term demand projections was the goal of this study, which served as the foundation for a more thorough investigation of the Moroccan energy market. This Fig. 1 shows the total net consumption of electricity and the total net consumption of renewable energy in Morocco (can be consulted at [11]).

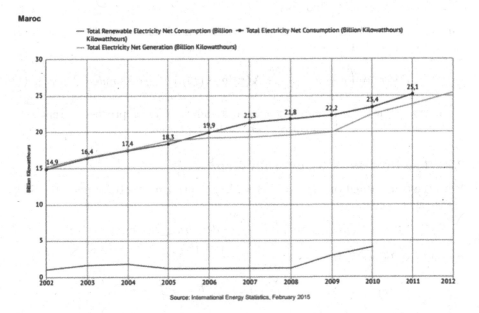

Fig. 1. The total net consumption of electricity and the total net consumption of renewable energy in Morocco

Numerous scientific disciplines employ the stochastic Gompertz diffusion process (SGDP) to simulate stochastic processes. The homogenous case of this process was first developed by Nobile et al. [1] in a theoretical form, and it was then used by Ferrante et al. [2] to study the proliferation of cancer cells, as well as by Gutiérrez et al. [3] to model Spanish GDP and CO_2 emissions using a bivariate stochastic Gompertz diffusion model and stock of automobiles in Spain [4]. However, the inhomogeneous case, with a constant deceleration coefficient and just the intrinsic growth rate (GR) in the drift touched by exogenous variables (functions of time and some parameters), was applied, for instance, to the cost of new homes in Spain [5] and the emission of CO2 [6]. Finally, Albano et. al. [7] studied a inhomogeneous terminology used to describe the impact of therapeutic interventions that might change theb. In the curent study, we define the SIGDP, it is employed in a variety of settings. Firstly we obtain the characteristics of probability of the process sush as their expression analytical, the transition probability density function (TPDF), and the mean functions (conditional and non-conditional). Then, Using a maximum likelihood (ML) technique, We make parameter estimates and getting confidence bounds. Finally, we used the SIGDP to examine the development of Morocco's electricity consumption in order to assess this process's capacity for modeling real data.

2   Basic Probabilistic Properties of the Model

2.1   The Suggested Process

The following diffusion process provides the model's stochastic counterpart $X(\tau); \tau \in [\tau_0; T]$ has values in $(0, \infty)$, $X(\tau)$ satisfies the SDE

$$dX(\tau) = \left(aX(\tau) - \frac{h'(\tau)}{h(\tau)} X(\tau) ln(X(\tau)) \right) dt + \sigma X(\tau) dw(\tau). \qquad (1)$$

where $\sigma > 0$, $w(\tau)$ is a one-dimensional standard Wiener process, a represent the intrinsic GR and the function $h(\tau)$ is differentiable.

2.2   The Process as Analytically Expressed

With the transformation $y(\tau) = h(\tau) \, ln(X(\tau))$, utilizing the Itô lemma, the SDE (1) leads

$$dy(\tau) = h(\tau) \left(a - \frac{\sigma^2}{2} \right) dt + \sigma h(\tau) dw(\tau),$$

By integrating, we have

$$y(\tau) = y(s) + (a - \frac{\sigma^2}{2}) \int_s^\tau h(\xi) \, d\xi + \sigma \int_s^\tau h(\xi) dw(\xi).$$

Finally, it follows that the explicit expression of the process's

$$X(\tau) = exp\left\{ \frac{h(s)}{h(\tau)} ln(X(s)) + \frac{(a - \frac{\sigma^2}{2})}{h(\tau)} \int_s^\tau h(\xi) \, d\xi + \frac{\sigma}{h(\tau)} \int_s^\tau h(\xi) dw(\xi) \right\} \qquad (2)$$

2.3 Distributed Probability

The random variable $\int_s^\tau h(\xi)dw(\xi)$ is normally distributed $\mathcal{N}_1(0, \int_s^\tau h^2(\xi)d\xi)$, so the random variable $X(\tau)$ is a one-dimensional lognormal process $X(\tau)/X(s) = X_s \sim \Lambda_1(\ (s, \tau, X_s), \sigma^2\nu^2(s, \tau))$, with

$$\lambda(s, \tau, X_s) = \frac{h(s)}{h(\tau)}ln(X(s)) + \frac{(a - \frac{\sigma^2}{2})}{h(\tau)}\int_s^t h(\xi)\ d\xi,$$

$$\nu^2(s, t) = \frac{1}{h^2(\tau)}\int_s^t h^2(\xi)d\xi.$$

The TPDF of this process $f(X, \tau | y, s)$ takes the form

$$f(X, \tau | y, s) = \frac{1}{X\sqrt{2\pi\sigma^2\nu^2(s, \tau)}}exp\left(-\frac{[ln(x) - \lambda(s, \tau, X)]^2}{2\sigma^2\nu^2(s, \tau)}\right). \qquad (3)$$

2.4 Computation

The r-th conditional moment of the process is determined by the features of the Lognormal distribution, and is given by

$$E(X^r/X(s) = X_s) = exp\left(r\lambda(s, \tau, X_s) + \frac{r^2\sigma^2\nu^2(s, \tau)}{2}\right).$$

For $r = 1$, the conditional mean function (CMF):

$$E(X(\tau)/X(s) = X_s) = exp\left\{\frac{h(s)}{h(\tau)}ln(X(s)) + \frac{\left(a - \frac{\sigma^2}{2}\right)}{h(\tau)}\int_s^\tau h(\xi)d\xi + \frac{\sigma^2}{2\,h^2(\tau)}\int_s^t h^2(\xi)d\xi\right\}. \qquad (4)$$

Considering the standard condition $P(X(\tau_1) = X_1) = 1$, the mean function (MF) takes the form:

$$E(X(\tau)) = exp\left\{\frac{h(\tau_1)}{h(\tau)}ln(X_{\tau_1}) + \frac{\left(a - \frac{\sigma^2}{2}\right)}{h(\tau)}\int_{\tau_1}^\tau h(\xi)d\xi + \frac{\sigma^2}{2\,h^2(\tau)}\int_{\tau_1}^\tau h^2(\xi)d\xi.\right\} \qquad (5)$$

3 Inference

In this part, we'll look at the ML estimate of the parameters from which we may get, by utilizing Zehna's theorem [8], corresponds to the parametric functions given above.

3.1 Parameter Estimation

From discrete observation $X_1, X_2, \cdots, X_n$ for times $\tau_1 < \tau_2 < \cdots < \tau_n$, with standard condition $P(X(\tau_1) = X_1) = 1$, the likelihood function can be expressed as

$$l(X_1, \cdots, X_n, \alpha, \sigma^2) = \prod_{i=2}^{n} f(X_i, \tau_i | X_{i-1}, \tau_{i-1}),$$

which is expressed as

$$l = \prod_{i=2}^{n} \frac{1}{X_i \sqrt{2\pi\sigma^2 \nu^2(s,\tau)}} exp\left(-\frac{\left\{ ln(X_i) - \frac{h(\tau_{i-1})}{h(\tau_i)} ln(X_{i-1}) - \frac{(a - \frac{\sigma^2}{2})}{h(\tau_i)} \int_{\tau_{i-1}}^{\tau_i} h(\xi)\, d\xi \right\}^2}{2\sigma^2 \nu^2(s,\tau)} \right).$$

We will present, this function drawn as a vector, taking into account the following transformation:

$$v_1 = X_1, \qquad and \qquad v_i = \nu_i^{-1}\left(ln(X_i) - \frac{h(\tau_{i-1})}{h(\tau_i)} ln(X_{i-1}) \right)$$

for $i = 2, \cdots, n$ with thus, this function takes the form

$$l(\mathbf{v}, \mathbf{a}, \sigma^2) = [2\pi\sigma^2]^{-(n-1)/2} exp\left\{ -\frac{1}{2\sigma^2}(\mathbf{v} - \mathbf{U}'\mathbf{a})'(\mathbf{v} - \mathbf{U}'\mathbf{a}) \right\}$$

at which

$$\mathbf{a} = a - \sigma^2/2, \ \mathbf{v} = (v_2, \cdots, v_n)',$$

$$\nu_i = \nu(\tau_{i-1}, \tau_i),$$

$$u_i = \frac{\nu_i^{-1}}{h(\tau_i)} \int_{\tau_{i-1}}^{\tau_i} h(\xi) d\xi$$

where the matrix $\mathbf{U} = (\mathbf{u}_2, \cdots, \mathbf{u}_n)$ is a line vector with $n - 1$ dimensions The ln-likelihood for Eq. (3) has the following form

$$\mathcal{L} = ln(l(\mathbf{v}, \mathbf{a}, \sigma^2)) = -\frac{n-1}{2} log(2\pi) - \frac{n-1}{2} log(\sigma^2) - \frac{1}{2\sigma^2}(\mathbf{v} - \mathbf{U}'\mathbf{a})'(\mathbf{v} - \mathbf{U}'\mathbf{a})$$

By deriving the $\mathcal{L}$ with respect to σ^2 and a we obtain

$$\frac{\partial \mathcal{L}}{\partial \sigma^2} = -\frac{n-1}{2\sigma^2} + \frac{1}{2\sigma^4}(\mathbf{v} - \mathbf{U}'\mathbf{a})'(\mathbf{v} - \mathbf{U}'\mathbf{a}) \tag{6}$$

$$\frac{\partial Log(l)}{\partial \mathbf{a}} = -\frac{1}{2\sigma^2} \frac{\partial[(\mathbf{v} - \mathbf{U}'\mathbf{a})'(\mathbf{v} - \mathbf{U}'\mathbf{a})]}{\partial \mathbf{a}} = \frac{1}{\sigma^2} \mathbf{U}(\mathbf{v} - \mathbf{U}'\mathbf{a}) \tag{7}$$

Making the derivatives (6) and (7) equal to zero, we obtain the following equations

$$- (n - 1)\sigma^2 + (\mathbf{v} - \mathbf{U}'\mathbf{a})'(\mathbf{v} - \mathbf{U}'\mathbf{a}) = 0 \tag{8}$$

$$\mathbf{U}\mathbf{v} - \mathbf{U}\mathbf{U}'\mathbf{a} = 0 \tag{9}$$

The Eqs. (8) and (9) becomes

$$\mathbf{U}\mathbf{v} = \mathbf{U}\mathbf{U}'\mathbf{a} \tag{10}$$

$$(n - 1)\sigma^2 = (\mathbf{v} - \mathbf{U}'\mathbf{a})'(\mathbf{v} - \mathbf{U}'\mathbf{a}) \tag{11}$$

The ML estimators of $\mathbf{a}$ and σ^2 yield

$$\hat{\mathbf{a}} = (\mathbf{U}\mathbf{U}')^{-1}\mathbf{U}\mathbf{v} \tag{12}$$

$$(n - 1)\hat{\sigma}^2 = \mathbf{v}'\mathbf{H}\mathbf{v} \tag{13}$$

where $\mathbf{H}$ is a symmetric, idempotent matrix that is provided by

$$\mathbf{H} = \mathbf{I}_{n-1} - \mathbf{U}'(\mathbf{U}\mathbf{U}')^{-1}\mathbf{U}.$$

3.2   Estimated Mean Functions

By changing the parameters in (4) and (5) with their estimators in (12) and (13), the Estimated Mean Function (EMF) and Estimated Conditional Mean Function (ECMF) of the proposed model are derived by using Zehna's theorem [1]. The ECMF also contains the following expressions:

$$E(X(\tau)) = exp\left\{\frac{h(s)}{h(\tau)}ln(X_s) + \frac{\left(\hat{\mathbf{a}} - \frac{\hat{\sigma}^2}{2}\right)}{h(\tau)}\int_s^\tau h(\xi)d\xi + \frac{\sigma^2}{2\,h^2(\tau)}\int_s^\tau h^2(\xi)d\xi.\right\} \tag{14}$$

In the standard condition $P(x(t_1) = x_1) = 1$, the EMF of the process is:

$$E(X(\tau)) = exp\left\{\frac{h(\tau_1)}{h(\tau)}ln(X_{\tau_1}) + \frac{\left(\hat{\mathbf{a}} - \frac{\hat{\sigma}^2}{2}\right)}{h(\tau)}\int_{\tau_1}^\tau h(\xi)d\xi + \frac{\sigma^2}{2\,h^2(\tau)}\int_{\tau_1}^\tau h^2(\xi)d\xi.\right\} \tag{15}$$

3.3   Properties of ML Estimators

Distribution of ML Estimators. The l may be rewritten as:

$$l(\mathbf{v}, \mathbf{a}, \sigma^2) = [2\pi]^{-\frac{(n-1)}{2}}|\sigma^2 I_{n-1}|^{-1/2}exp\left\{-\frac{1}{2}(\mathbf{v} - \mathbf{U}'\mathbf{a})'(\sigma^2 I_{n-1})^{-1}(\mathbf{v} - \mathbf{U}'\mathbf{a})\right\}$$

where we conclude that

$$\mathbf{v} \sim \mathcal{N}_{n-1}(\mathbf{U}'\mathbf{a}, \sigma^2 I_{n-1})$$

then

$$(\mathbf{U}\mathbf{U}')^{-1}\mathbf{U}\mathbf{v} \sim \mathcal{N}_1\left((\mathbf{U}\mathbf{U}')^{-1}\mathbf{U}\mathbf{U}'\mathbf{a}, \sigma^2(\mathbf{U}\mathbf{U}')^{-1}(\mathbf{U}\mathbf{U}')(\mathbf{U}\mathbf{U}')^{-1}\right)$$

and therefore, we have

$$\hat{\mathbf{a}} \sim \mathcal{N}_1\left(\mathbf{a}, \sigma^2(\mathbf{U}\mathbf{U}')^{-1}\right)$$

The distribution of $\hat{\sigma}^2$, can be obtained from the corollary 2.11.2 in [9]:
If $Z \sim \mathcal{N}_q[\lambda, \sigma]$, σ non singular and $\mathbf{B}_{q \times q}$ symmetric, then, $Z'\mathbf{B}Z \sim \chi_k^2(\gamma)$, where $k = rank(\mathbf{B})$ and $\gamma = \lambda'\mathbf{B}\lambda$ IF, ONLY IF $\mathbf{B}\sigma$ is idempotent. Due to $\mathbf{H}$'s symmetry and idempotence, then,

$$rank(\mathbf{H}) = tr(\mathbf{H}_U) = n - 2,$$

then applying the latest consequent particularly in this case: $Z = \sigma^{-1}\mathbf{v}$, $\sigma = I_{n-1}$, $\mathbf{B} = \mathbf{H}$ and $\lambda = \mathbf{U}'\mathbf{a}$, we obtain

$$\frac{\mathbf{v}'}{\sigma}\mathbf{H}\frac{\mathbf{v}}{\sigma} \sim \chi_{n-2}^2(\delta), \ with \ \gamma = \mathbf{a}'\mathbf{U}\mathbf{H}\mathbf{U}'\mathbf{a} = 0$$

and therefore

$$\frac{(n-1)\hat{\sigma}^2}{\sigma^2} \sim \chi_{n-2}^2$$

The independence between $\hat{\mathbf{a}}$ and $\hat{\sigma}^2$ is substantiated by using the corollary 2.11.4, p.66 in the [9]:
Let $Z \sim \mathcal{N}_q[\lambda, \sigma]$, with $\sigma > 0$. Then, $y_j = Z'A_j Z + 2b_j'Z + c_j$, $j = 1, 2$ are independent distribution IF, ONLY IF $A_1\sigma A_2 = 0$, $A_2\sigma b_1 = 0$, $A_1\sigma b_2 = 0$, and $b_1'\sigma b_2 = 0$.
If we choose $Z = \mathbf{v} \sim \mathcal{N}_{n-1}(\mathbf{U}'\mathbf{a}, \sigma^2 I_{n-1})$; $A_1 = \mathbf{H}$; $b_1 = 0$; $c_1 = 0$ and $A_2 = 0$; $b_2 = (\mathbf{U}\mathbf{U}')^{-1}\mathbf{U}$ and $c_2 = 0$ therefore the preceding corollary's necessary and sufficient criteria are satisfied, and as a result $(\mathbf{U}\mathbf{U}')^{-1}\mathbf{U}\mathbf{v}$ and $\mathbf{v}'\mathbf{H}\mathbf{v}$ are independently distributed, which means that $\hat{a}$ and $\hat{\sigma}^2$ as well.

3.4   The Estimators' Completeness and Sufficiency

The function l becomes

$$l(\mathbf{v}, \mathbf{a}, \sigma^2) = \frac{1}{[2\pi\sigma^2]^{\frac{(n-1)}{2}}} exp\left\{-\frac{1}{2\sigma^2}[(n-1)\hat{\sigma}^2 + (\hat{\mathbf{a}} - \mathbf{a})'\mathbf{U}\mathbf{U}'(\hat{\mathbf{a}} - \mathbf{a})]\right\}$$

from which, we can conclude that $(\hat{\mathbf{a}}, \hat{\sigma}^2)$ is an exhaustive statistic for $(\mathbf{a}, \sigma^2)$. Similar to the methodology described for the ML estimators of the multivariate normal distribution parameters outlined in Anderson [10].
Then $\hat{\mathbf{a}}$ and $\frac{(n-1)\hat{\sigma}^2}{(n-2)\sigma^2}$ are the UMVUE for $\mathbf{a}$, σ^2, respectively.

3.5   Confidence Bounds

Using the method outlined in [12] we obtain the confidence bounds for the parameter σ^2. The random variable z given by

$$z = \frac{ln(\frac{X(\tau)}{X(s)}) - \lambda(s,\tau,X_s)}{\sigma\sqrt{\tau - s}} \sim N(0,1)$$

A $100(1 - k)\%$ confidence bound for z is given by $P(|z| \leq \delta) = 1 - k$. We may get confidence bounds of the following type from this:

$$X_l(\tau) = X_s exp[\lambda(s,\tau,X_s) - \delta\sigma\sqrt{\tau - s}] \tag{16}$$

$$X_u(\tau) = X_s exp[\lambda(s,\tau,X_s) + \delta\sigma\sqrt{\tau - s}] \tag{17}$$

with $\delta = \psi^{-1}(1 - \frac{k}{2})$ and where ψ denote the inverse cumulative normal standard distribution. Then, by substituting the parameters by their estimators in (16) and (17), the estimated confidence bounds are given by

$$\hat{X}_l(\tau) = X_s exp[\hat{\lambda}(s,\tau,X_s) - \delta\hat{\sigma}\sqrt{\tau - s}] \tag{18}$$

$$\hat{X}_l(\tau) = X_s exp[\hat{\lambda}(s,\tau,X_s) + \delta\hat{\sigma}\sqrt{\tau - s}] \tag{19}$$

where $\hat{\lambda}(s,\tau,X_s) = \frac{h(s)}{h(\tau)}ln(X(s)) + \frac{(\hat{a} - \frac{\hat{\sigma}^2}{2})}{h(\tau)}\int_s^\tau h(\xi) \, d\xi,$

4   Application

The proposed model was applied to actual data for Morocco's total electric power usage (reported in billion kilowatts) from 1980 to 2012. These statistics, which related to sales by ONE, the Moroccan authority, are accessible at [11]. Two steps that make up the methodology are as follows:
• The first step: To estimate the model's parameters, start with the first 30 data in the sequence of observations being analyzed, using expressions (12) and (13). Then establish the relevant confidence bounds using Eqs. (??) and (??).
• The second step: Forcast the associated values for Morocco's electric power consumption for the years 2011 and 2012 using the EMF and ECMF in (14) and (15), and then contrast the results with the corresponding ral data.

For the computations needed for the present study, a Matlab application was used. Think about it, for example, the function $h(\tau) = \frac{1 - t^2 - t^4}{t + 1}$, The corresponding estimators' values, and the confidence bounds, are $\hat{a} = 0.060651$ and $\hat{\sigma} = 1.094854.10^{-3}$ with confidence bounds $(0.048313; 0.072988)$ and $(0.699151; 1.956171).10^{-3}$.

Table 1 explains the fit and forecast made possible by the EMF with its EMF_l and EMF_u.
Table 2 explains the fit and forecast made possible by the $ECMF$ with its $ECMF_l$ and $ECMF_u$.

Table 1. Real data, EMF, EMF_l and EMF_u

Year	Real data	EMF	EMF_l	EMF_u
1980	4.409	4.409	4.409	4.409
1981	4.774	4.676	4.615	4.734
1982	5.130	4.959	4.832	5.082
1983	5.612	5.259	5.057	5.455
1984	5.776	5.577	5.294	5.856
1985	5.884	5.914	5.541	6.284
1986	6.568	6.269	5.798	6.744
1987	7.018	6.646	6.068	7.237
1988	7.656	7.045	6.349	7.764
1989	7.744	7.467	6.643	8.329
1990	8.370	7.914	6.951	8.935
1991	8.877	8.387	7.272	9.583
1992	9.804	8.888	7.607	10.277
1993	10.218	9.417	7.957	11.021
1994	10.350	9.977	8.323	11.817
1995	11.404	10.569	8.706	12.669
1996	11.617	11.196	9.105	13.581
1997	12.114	11.859	9.521	14.558
1998	12.935	12.560	9.956	15.603
1999	13.103	13.301	10.411	16.721
2000	13.050	14.085	10.885	17.917
2001	14.351	14.914	11.380	19.198
2002	14.856	15.790	11.897	20.567
2003	16.361	16.716	12.437	22.032
2004	17.411	17.695	13.001	23.599
2005	18.315	18.730	13.588	25.275
2006	19.872	19.824	14.202	27.067
2007	21.266	20.980	14.842	28.983
2008	21.751	22.201	15.511	31.032
2009	22.243	23.492	16.208	33.222
2010	24.844	24.855	16.935	35.563
2011	26.871	26.296	17.695	38.065
2012	28.946	27.818	18.486	40.739

Table 2. Real data, ECMF, $ECMF_l$ and $ECMF_u$

Year	Real data	ECMF	$ECMF_l$	$ECMF_u$
1980	4.409	4.409	4.409	4.409
1981	4.774	4.676	4.616	4.732
1982	5.130	5.063	4.998	5.123
1983	5.612	5.440	5.370	5.504
1984	5.776	5.950	5.874	6.021
1985	5.884	6.123	6.045	6.196
1986	6.568	6.238	6.158	6.312
1987	7.018	6.962	6.873	7.045
1988	7.656	7.438	7.343	7.527
1989	7.744	8.113	8.010	8.210
1990	8.370	8.206	8.102	8.304
1991	8.877	8.869	8.756	8.974
1992	9.804	9.405	9.285	9.517
1993	10.218	10.386	10.254	10.509
1994	10.350	10.824	10.686	10.953
1995	11.404	10.964	10.824	11.094
1996	11.617	12.078	11.924	12.222
1997	12.114	12.304	12.147	12.449
1998	12.935	12.829	12.666	12.982
1999	13.103	13.697	13.523	13.861
2000	13.050	13.875	13.698	14.041
2001	14.351	13.819	13.643	13.983
2002	14.856	15.195	15.001	15.375
2003	16.361	15.729	15.528	15.916
2004	17.411	17.319	17.098	17.525
2005	18.315	18.429	18.193	18.648
2006	19.872	19.385	19.137	19.6147
2007	21.266	21.031	20.762	21.281
2008	21.751	22.503	22.215	22.770
2009	22.243	23.016	22.722	23.289
2010	24.844	23.536	23.235	23.815
2011	26.871	26.284	25.948	26.596
2012	28.946	28.426	28.062	28.763

Figure 2 illustrates the comparison of the real data with the EMF, EMF_l and EMF_u.

3 illustrates the fit and forecasting made with the $ECMF$ of the model with respect to real data.

MATLAB was used to do all computations.

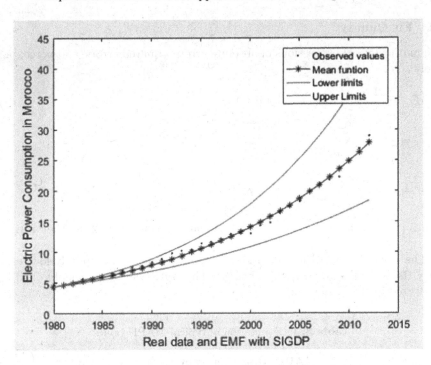

Fig. 2. Illustrates the comparison of the real data with EMF, EMF_l and EMF_u

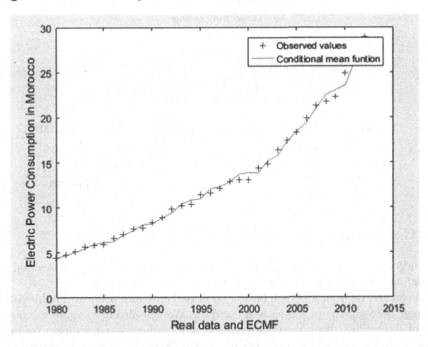

Fig. 3. Illustrates the fit and forecasting made with the $ECMF$ of the model with respect to real data

4.1   Fit Quality

The mathematical expressions of absolute errors, squared errors, and percentage errors:

$$MAE \ = \frac{1}{N} \sum_{i=1}^{n} \mid x(t_i) - \hat{x}(t_i) \mid,$$

$$RMSE = \sqrt{\frac{1}{N} \sum_{i=1}^{n} (x(t_i) - \hat{x}(t_i))^2},$$

$$MAPE = \frac{1}{N} \sum_{i=1}^{n} \frac{\mid x(t_i) - \hat{x}(t_i) \mid}{x(t_i)} \times 100.$$

with $\hat{X}(\tau)$ is obtained by substituting the parameters in Eq. (2) by their estimators.

The values obtained for the above error measures are acceptably low, especially the MAPE according to Table (3). The statistical measures obtained are illustrated in the Table (4).

Table 3. Interpretation of typical MAPE values.

MAPE	Interpretation
< 10	Highly accurate forecasting
20 − 30	Good forecasting
30 − 50	Reasonable forecasting
> 50	Inaccurate forecasting

Table 4. Fit quality of the model.

Measures of Forecasting Accuracy Error	Values of SIGDP
MAE	0.361247990035216
RMSE	0.459024822910505
MAPE	2.759618319944835

5   Conclusions

A study of the SIGDF, including all of its probabilistic characteristics and the accompanying statistical inference, is presented in this article.

In the future, these models will be able to fit real data and produce goodness of fit findings between the processes and the data. We will also investigate the potential of describing all of these processes in their inhomogeneous form by introducing exogenous variables and examining the usage of numerical approaches to acquire estimates.

Abbreviations

The following acronyms are utilized in this document:

GR: Growth Rate
SGDP: Stochastic Gompertz Diffusion Process
GDP: Gross Domestic Product
SIGDP: Stochastic Inhomogeneous Gompertz Diffusion Process
TPDF: Transition Probability Density Function
SDE: Stochastic Differential Equation
CMF: Conditional Mean Function
ML: Maximum Likelihood
EMF: Estimated Mean Function
ECMF: Estimated Conditional Mean Function
MAE: Mean Absolute Error
RMSE: Root Mean Square Error
MAPE: Mean Absolute Percentage Error
UMVUE: Minimum variance unbiased estimator

References

1. Nobile, A.G., Ricciard, L.M., Sacerdote, L.: On Gompertz growth model and related difference equations. Biol. Cybern. **42**(3), 221–229 (1982)
2. Ferrante, L., Bompadre, S., Possati, L., Leone, L.: Parameter estimation in a Gompertzian stochastic model for tumor growth. Biometrics **56**(4), 1076–1081 (2000)
3. Gutiérrez, R., Gutiérrez-Sánchez, R., Nafidi, A.: A bivariate stochastic Gompertz diffusion model: statistical aspects and application to the joint modeling of the Gross Domestic Product and CO_2 emissions in Spain. Environmetrics **19**(6), 643–658 (2005)
4. Gutiérrez, R., Gutiérrez-Sánchez, R., Nafidi, A.: Modelling and forecasting vehicle stocks using the trends of stochastic Gompertz diffusion models, the case of Spain. Appl. Stoch. Model. Bus. Ind. **25**(3), 385–405 (2009)
5. Gutiérrez, R., Gutiérrez-Sánchez, R., Nafidi, A., Román, P., Torres, F.: Inference in Gompertz-Type nonhomogeneous stochastic systems by means of discrete sampling. Cybern. Syst. **36**(2), 203–216 (2005)
6. Gutiérrez, R., Gutiérrez-Sánchez, R., Nafidi, A.: Trend analysis using non-homogeneous stochastic diffusion processes. Emission of CO2; Kyoto protocol in Spain. Stochastic Environ. Res. Risk Assess. **22**(1), 57–66 (2008)
7. Albano, G., Giorno, V., Román-Román, P., Román-Román, S., Torres-Ruiz, F.: Estimating and determining the effect of a therapy on tumor dynamics by means of a modified Gompertz diffusion process. J. Theor. Biol. **36**(4), 206–219 (2015)
8. Zehna, P.W.: Invariance of maximum likelihood estimators. Ann. Math. Stat. **37**(3), 744 (1966)
9. Srivastava, M.S., Khatri, C.G.: An Introduction to Multivariate Statistics. North Holland, New York (1979)
10. Anderson, T.W.: An introduction to multivariate statistical analysis, 2nd edn. Wiley, New York (1984)

11. MAROC portail de données. https://morocco.opendataforafrica.org/. Accessed Feb 2015
12. Katsamaki, A., Skiadas, C.: Analytic solution and estimation of parameters on a stochastic exponential model for technology diffusion process. Appl. Stoch. Models Data Anal. **11**(1), 59–75 (1995)

Localization and Navigation of ROS-Based Autonomous Robot in Hospital Environment

Hamza Ben Roummane[1]([✉]) [iD] and Cherki Daoui[2]

[1] Departement of Computer Science, Sultan Moulay Slimane University, Beni Mellal, Morocco
hamza.benroummane@gmail.com
[2] Departement of Mathematics, Sultan Moulay Slimane University, Beni Mellal, Morocco

Abstract. This work is part of a research project during the COVID-19 pandemic that aims to design and develop a mobile autonomous robot for hospitals. In practice, implementing a navigation program directly on a physical robot is both expensive and hazardous. The solution is to perform a simulation using ROS (Robot Operating System), which offers several advantages that make it an appealing option for testing and development. In an unknown hospital environment, this paper presents a simulation of the navigation process of the autonomous robot Turtlebot3 by employing the Simultaneous Localization and Mapping (SLAM) algorithm, specifically the GMapping method, utilizing the distributed software framework of ROS. In a known hospital environment, we utilize trajectory planning algorithms designed for deterministic models. However, considering the inherent uncertainty in the environment and the inevitable inaccuracies of the models, we integrate the Markov decision process (MDP) by applying the classical Q-Learning algorithm. Through these simulations, our aim is to test and refine the navigation algorithms to enhance the performance of our mobile robot. Ultimately, the proposed simulation approach contributes to the development of robotic solutions that can assist in performing various routine tasks remotely. This saves time for healthcare personnel and, most importantly, ensures their safety.

Keywords: ROS · SLAM · GMapping algorithm · Turtlebot3 robot · MDP · Simulation · Q-Learning algorithm

1 Introduction

Robotics has become an integral part of our daily lives, improving efficiency and making tasks easier and more convenient through its integration into various industries and household activities. The field has brought about advancements in medicine, where robots assist in surgeries, and education, where they provide interactive learning experiences. Additionally, robotics has created new job opportunities and industries. It is clear that robotics plays a significant role in shaping and improving our daily lives and will continue to do so in the future [1].

The primary challenge in mobile robotics today is the development of intelligent navigation systems. Autonomous navigation is a research area aimed at giving machines the ability to move in an environment without human assistance or intervention to achieve a specific goal [2]. The navigation function provides the robot with the information it needs to make decisions independently and equips it with the means of locomotion that are best suited for its environment. The robot must choose the best actions to complete its mission [3].

The ROS navigation package incorporates reinforcement learning and MDP (Probabilistic Decision Making) algorithms, which allow autonomous robots to navigate environments without human intervention. Reinforcement learning involves training an agent to make decisions based on feedback received from the environment, while MDP involves making probabilistic estimates of the consequences of different actions and choosing the action with the highest expected utility [4]. These algorithms are used to teach robots how to navigate in different environments, making decisions about which path to take or which action to perform based on the information received from their sensors and the desired outcome. By using these algorithms, ROS-based autonomous robots can perform complex tasks in a variety of settings, including hospital environments.

In practice, implementing a navigation program directly on a physical robot is both expensive and dangerous. The solution is to perform a simulation using ROS (Robot Operating System), which provides several advantages that make it an attractive option for testing and development. Firstly, simulation offers a cost-effective solution as it eliminates the need for physical robots, reducing expenses. Secondly, it provides a safe testing environment, as simulations can be run in a controlled setting, making it easier to test complex or hazardous scenarios that might be difficult or dangerous to perform in the real world. Finally, simulation helps to improve the accuracy of the navigation system by allowing for the refinement of algorithms and parameters, leading to better performance. Overall, simulating autonomous navigation robots with ROS provides a practical, safe, and efficient way to develop and test navigation systems [5].

As humans often, learn about the world through their senses. Without certain sensory sources to obtain information about the environment, a robot cannot explore an uncharted area [6]. Odometers, laser rangefinders, GPS, inertial measurement units (IMUs), navigation and sound ranging systems (Sonar), and cameras are just a few of the many types of sensors used to enable robots to detect various situations [7]. The simulated robot is Turtlebot3, which is an open-source robot for education and research, with features such as navigation and object recognition through sensors and actuators with a compact size and good price [8].

This project has been motivated by the need for safe and efficient navigation of autonomous robots in hospital environments, especially during the COVID-19 pandemic. The objective of this project is to develop and test a navigation system for Turtlebot3 using ROS in a simulated hospital environment. The system aims to aid healthcare professionals in tasks such as delivery of medical supplies, disinfection of hospital wards, and patient monitoring, while minimizing the risk of viral transmission. The simulation model will enable the testing and refinement of the navigation algorithms, as well as the evaluation of the system's performance in different hospital scenarios. Overall, the project seeks to contribute to the development of robotic solutions that can support healthcare workers in providing safe and effective care to patients.

To realize an intelligent navigation system, this paper aims to build a simulation model of the navigation in Hospital environment of the autonomous robot Turtlebot3 with ROS. First, we describe the use of the ROS navigation package and the tools needed to implement it, and then we elaborate the necessary configuration and steps to perform the simulation. The study process will be divided in two different situations, first, the robot is in an unknown Hospital environment, so it is necessary to use a SLAM algorithm, which allows both the robot navigation and localization. Then, in the second case, the environment is known, that is, the map will be among the input elements in the algorithm.

The essay is organized in the following manner. Section 2 provides an overview of related research in the field of localization and navigation of autonomous robots. Section 3 introduces the history and concepts of ROS. Section 4 presents the experience of navigating in an unknown environment. Section 5 discusses navigation in a known environment. The results and discussions are presented in Sect. 6. The essay concludes in Sect. 7.

2 Related Work

Previous studies have focused on the development of mobile robots and their implementation in simulation environments. Denis Chikurtev [9] presented the creation of a mobile robot model in the Gazebo simulation environment, including the configuration options and functionalities for autonomous navigation. While Chikurtev's work emphasizes the simulation environment, practical validation in real-world scenarios is lacking.

Arbnor Pajaziti et al. [10] described the ROS-based navigation and mapping system of the Turtlebot robot, showcasing its effectiveness in navigation tasks. However, further exploration is needed to evaluate the system's performance and robustness beyond the scope of their study. To address the high cost and complex structure of autonomous navigation robots, Zhang, Huijuan, et al. [11] proposed a low-cost and simple scheme for mobile robot placement and navigation in interior environments, which was experimentally confirmed. Although their scheme shows promise, its applicability in different contexts and environments requires further investigation. Additionally, Alessandro Bianco et al. [12] discussed ongoing efforts to develop an affordable, off-the-shelf, and fully autonomous ground rover platform for monitoring and intervention activities on farms, utilizing the Robot Operating System (ROS). While their work provides valuable insights into the practical implementation of autonomous systems, it lacks a comprehensive integration of Simultaneous Localization and Mapping (SLAM) algorithms and Markov Decision Processes (MDP) for enhanced autonomy.

These previous works have provided valuable insights and inspiration for the current study. However, they also have certain limitations. Practical validation in real-world scenarios, evaluation of system performance and robustness beyond specific settings, and comprehensive integration of SLAM algorithms and MDP are areas that need further exploration.

In this study, we aim to bridge these gaps and contribute to the field of autonomous robot navigation. Our research plan involves integrating SLAM algorithms and MDP to enable effective planning and execution of robot trajectories in diverse environments, including hospitals. Our solution harnesses the potential of ROS by overcoming identified limitations, presenting an avenue for improved safety and efficiency in healthcare delivery systems. By advancing the recent body of research, we present an overall framework that considers previous limitations, culminating in significant gains in autonomous robot navigation.

3 Introduction to ROS

Before the advent of the Robot OS, every robotics researcher spent a considerable amount of time designing the hardware of their robot and the associated embedded software. This required skills in mechanics, electronics and embedded programming [13]. Hence, the need for the Robot Operating System (ROS), which provides a standard framework for creating robotic applications, making it attractive for use with robots. By abstracting away many of the low-level complexities and allowing developers to focus on the high-level logic of their robots, ROS provides a collection of libraries and tools that simplify the creation of code for robots. In addition, ROS offers a large ecosystem of community-developed packages and tools, which simplifies the construction and integration of various components of a robotic system. This can result in faster development of robotic projects and higher production [14].

3.1 Programming with ROS

The Robot Operating System (ROS) operates in a peer-to-peer fashion, enabling nodes to communicate directly with each other, eliminating the requirement for a central server. ROS' support for multiple programming languages, such as C++ and Python, makes it available to a broad spectrum of developers. Furthermore, being open-source software, developers have the ability to access the source code, take part in its development, and enjoy the advantages of a thriving developer community [15].

3.2 Tools of ROS

A set of tools is provided by ROS to assist developers in build, debug and maintaining robotic systems. ROS offers a variety of tools and libraries, some of which are:

- ROSCORE: The ROS Master, which is the central repository for all ROS topics, services, and parameters.
- ROSRUN and ROSLAUNCH: Command-line tools for launching ROS nodes and managing ROS processes.
- ROSNODE and ROSTOPIC: Tools for introspecting and debugging ROS nodes and topics.
- RQT: A platform that manages the deployment of Qt-based plugins. Some basic plugins are provided by ROS and it is also possible to develop custom modules.
- RVIZ: is a 3D visualization tool for the Robot Operating System (ROS). It is used to visualize various data from ROS topics, such as sensor data, odometry information, and robot model information.

- GAZEBO: is a multi-robot simulator for the Robot Operating System (ROS). It provides a virtual environment for testing and developing robot applications, and allows for testing in realistic and complex scenarios that would be difficult or impossible to replicate in the real world.

These are some of the most widely used tools in ROS, but the framework provides many more tools and libraries to help developers build, debug, and maintain complex robotic systems [16].

3.3 ROS-Compatible Robots

Many robots are compatible with (ROS). Some of them are:

- TurtleBot: A small, low-cost robot designed for educational and research purposes [17].
- PR2: A humanoid robot developed by Willow Garage, which is widely used in research and academia [18].
- Clearpath Robotics Husky: A rugged and versatile outdoor robot, designed for industrial and research applications [19].
- KUKA LBR iiwa: A lightweight, high-precision industrial robot, widely used in manufacturing and assembly applications [20].
- ABB YuMi: A collaborative robot designed for industrial assembly application [21].

4   Using the Simultaneous Localization and Mapping

Simultaneous Localization and Mapping (SLAM) is a widely used computational approach in robotics for building maps of unknown environments and simultaneously determining the location of the robot within that environment [22]. The SLAM algorithm leverages sensor data obtained from the robot, such as images, lidar scans, or other measurements, to construct a map of the environment over time. Additionally, the algorithm estimates the robot's position and orientation relative to the map by comparing sensor data with previously mapped features [23]. There exist various SLAM algorithm variations, including EKF-SLAM, FastSLAM, and GraphSLAM, each with its own unique approach to estimating both the robot's pose and the map. SLAM algorithms are employed in a wide range of applications, including autonomous vehicles, drones, and robotic exploration of unknown environments, as well as augmented and virtual reality [24].

TurtleBot and its successors have a well-established SLAM feature, and TurtleBot3 is capable of creating a map using its compact and cost-effective platform [25]. In experimental settings where the robot lacks prior knowledge of the environment, it is required to navigate and map the area simultaneously, a process that necessitates concurrent navigation and localization. In addition, this task involves collecting sensor data and processing it to create a metric map, which is then stored in the robot's memory for later use in navigation.

All commands used in this work are specified in ROS documentation status. The first step is to display the environment one more time, the command below calls the environment as a 3D model.

```
roslanch turtlebot3_gazebo turtlebot3_house.launch
```

The next command presents the call to SLAM algorithms it is necessary to choose the desired SLAM method. We used Gmapping ho is a popular open-source software for SLAM and is based on the FastSLAM algorithm, which uses a particle filter to estimate a robot's pose (position and orientation) and feature locations in the environment [26]. The resulting map is a grid-based representation of the environment shown in Fig. 1 and 2, with each cell containing information about the probability that it is occupied or unoccupied. The map can be used for localization, path planning and obstacle avoidance.

```
roslanch turtlebot3_slam turtlebot3_slam.launch
slam_methods:=gmapping slam_methods:=gmapping
```

The drawn part of the map is the area of occupation of the robot sensors that we need to navigate with the robot so that it can perform the mapping operation. The following command allows us to know the navigation keys in our machine.

```
roslanch turtlebot3_teleop turtlbot3_teleop_key.launch
```

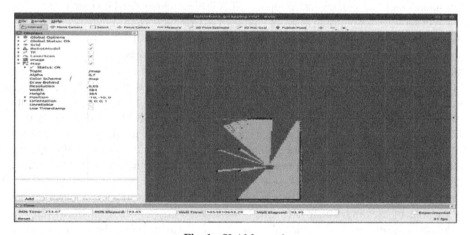

Fig. 1. SLAM map 1

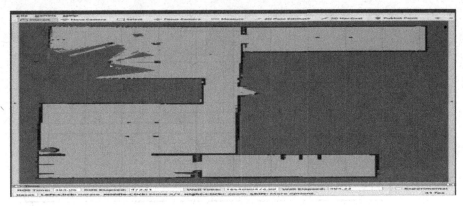

Fig. 2. SLAM map 2

It is necessary to make a change in the navigation parameters for an optimal and efficient path to the goal. As output of the command we have an interface rqt_reconfigure_param, the parameters are default. In this part we are interested in the three parameters framed in black in Fig. 3 {patch_distance_bias, goal_distance_bias, occdist_scale}. In the case where the parameters are default the robot follows the path, but with an offset that is due to goal_distance_bias not equaling zero. To increase the efficiency and decrease the time for the robot to reach the goal it is necessary to minimize path_distance_bias and occdist_scale and maximize goal_distance _bias. To view the robot parameters, simply call the following commands:

```
rosrun rqt_reconfigure rqt_reconfigure
```

The interface rqt_reconfigure_param is depicted in Fig. 3.

5 Navigation in a Known Environment

5.1 Navigation Using Path Planning Algorithms

The present study outlines the initial steps taken to simulate a known environment using TurtleBot3. Installation of the TurtleBot3 simulation package was carried out, with prerequisites including TurtleBot3 and TurtleBot3_MSGS. The robot was then positioned in a metric map recorded during the simultaneous localization and mapping (SLAM) stage, and displayed in a grid environment in Rviz. Navigation algorithms were applied to aid in movement within the environment. Orientation and location of the robot were obtained using the ROS Topic command in relation to a landmark. Subsequently, the robot was observed to be surrounded by multiple colors. Figure 4 show the Turtlebot3 within the Rvis map, surrounded by the key components of the navigation environment. These components include the global planner, local planner, laser scan, and AMCL orientation. These elements are crucial for enabling the robot to navigate autonomously

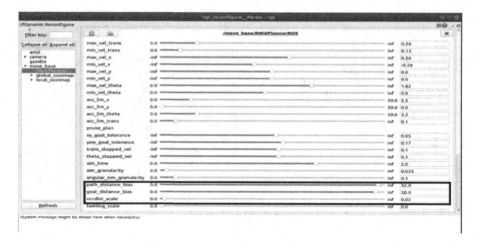

Fig. 3. rqt_reconfigure_param interface

within the hospital environment. The image demonstrates how these various components interact to support the robot's localization and planning capabilities.

Fig. 4. The main components of navigation environment

We focus on two planners in this study:

Global planner: In our approach, we employ a variety of global planners within the navigation stack [27] that adhere to the specified interface. One such planner is the carrot_planner, which provides a simplistic strategy. It verifies if the designated goal is obstructed and, if so, selects an alternative goal in close proximity to the initial target. However, in complex indoor environments, the practicality of this planner is limited.

Another type of global planner utilized is the navfn, which employs the Dijkstra algorithm to determine the global path with the minimum cost between the starting point and the desired endpoint. This algorithm efficiently searches for an optimal route based on the given environment.

Additionally, we utilize a more flexible global planner known as the global planner. It serves as a versatile replacement for the navfn planner and offers additional options to customize the path planning process. These options include support for the A* algorithm, the ability to toggle quadratic approximation, and the grid path toggle [28].

To configure the global planner in our simulation, we have employed the following settings:

- allow_unknown: true
- use_dijkstra: true
- use_quadratic: true
- use_grid_path: false

These settings enable us to handle various scenarios and adapt the global planning approach to suit our specific needs. By incorporating these different global planners and their configurations, our system can efficiently navigate in complex environments while providing flexibility and customization options.

The quality of the globally planned path is determined by three key parameters: a cost factor, neutral cost, and lethal cost, which collectively influence the path selection and evaluation process:

$$cost = NEUTRAL\ COST + COST\ FACTOR * costmap\_cost\_value \qquad (1)$$

Experimental findings have corroborated the aforementioned explanation, highlighting the significance of appropriately configuring the cost factor. Suboptimal outcomes are observed when the cost factor is set too low or too high, resulting in paths that lack sufficient clearance around obstacles and exhibit relatively flat curvatures. Through iterative experimentation, it was determined that a cost factor of 0.55, a neutral cost of 66, and a lethal cost of 253 yield highly desirable global paths.

Regarding the local planner, the nav_core::BaseLocalPlanner interface encompasses three distinct implementations: the dwa_local_planner, eband_local_planner, and teb_local_planner [29]. Each of these planners utilizes distinct algorithms to generate velocity commands. Specifically, the dwa_local_planner employs the dynamic window approach (DWA), generating a (v;w) pair that represents an optimal circular trajectory given the robot's local conditions. This is achieved through a comprehensive search of the velocity space within the subsequent time interval [30].

Finally, after adjusting the navigation parameters and algorithms, it is sufficient to indicate the goal in which the robot will perform its task to result in secure navigation away from obstacles detected by the robot's sensors. This method ensures that the robot navigates to its goal or performs a task, such as in the case of a vacuum cleaning robot. Figure 5 depicts the Turtlebot3 with a path plotted towards the final goal. The path is visualized, leading from the robot's initial position to the destination point. Figure 6 illustrates the Turtlebot3 successfully reaching its final goal.

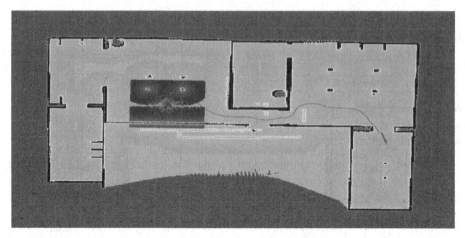

Fig. 5. The path of the robot to the goal

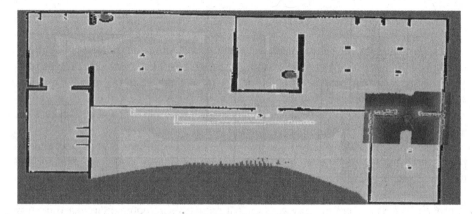

Fig. 6. Final station of the robot

5.2 Navigation Using MDP

Working with A*, Dijkstra, and other types of path planning algorithms has posed several problems, among them being the changing environment. Since our project involves a hospital robot, the environment will be dynamic. Therefore, in order to tackle these challenges, we propose incorporating the Markov Decision Process (MDP) and, specifically, the Q-learning algorithm. Q-learning falls within the scope of MDP and is a reinforcement learning algorithm that utilizes dynamic programming methods to learn an optimal policy by maximizing a cumulative reward function.

By leveraging MDP and reinforcement learning (RL) techniques, our approach offers several advantages. Firstly, it enables adaptability, allowing the Turtlebot 3 to adjust to changes in the environment. As the hospital environment may have varying obstacles, the robot can learn and update its navigation strategy based on the rewards obtained from different actions. Secondly, MDP takes into account uncertainty, enabling the robot to generate more precise trajectories by considering probabilistic factors associated with obstacles and sensor measurements. This capability enhances the robot's ability to navigate safely and efficiently in dynamic environments.

Furthermore, MDP and RL facilitate knowledge generalization. Once the Turtlebot 3 has learned optimal navigation policies in one environment, it can apply the acquired knowledge to similar environments. This transfer of knowledge reduces the need for extensive retraining or recalibration in each specific setting, saving time and effort during deployment.

By employing the Q-learning algorithm within the framework of MDP, our proposed approach addresses the limitations of traditional path planning algorithms and offers a more robust and adaptive solution for autonomous navigation in dynamic hospital environments. The use of MDP and RL techniques not only enhances the robot's adaptability and trajectory planning but also allows for the generalization of knowledge, making the system efficient and scalable for different hospital scenarios.

To implement Q-learning in ROS using the Turtlebot3 robot, you simply need to install the "turtlebot3_rl" package. This package provides scripts and configuration files for reinforcement learning on the Turtlebot3 robot using the following command:

```
sudo apt-get install ros-melodic-turtlebot3-rl
```

After configuring the virtual environment in GAZEBO in the same way as the other experiments, we created a ROS node to run the Q-learning algorithm using the "rl" package in ROS to implement Q-learning. The reinforcement learning package in ROS provides tools to implement reinforcement learning algorithms like Q-learning, which uses a Q table that is iteratively updated based on the current state and action chosen by the robot, storing the utility values for each possible state-action pair. This enables

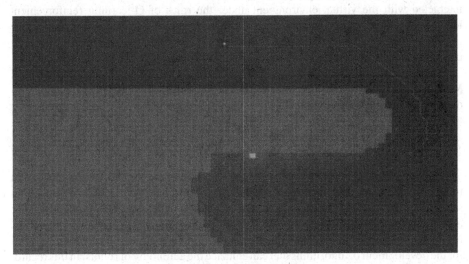

Fig. 7. Navigation with MDP

Turtlebot3 to make optimal decisions by selecting the action with the highest Q value in a given state (Fig. 7).

6 Experimental Results and Analysis

The results of the tests and analyses conducted in this study demonstrate the effectiveness of using SLAM in TurtleBot3 navigation in unknown environments using the Gmapping technique. The grid map produced by Gmapping offers a reliable representation of the environment as well as information on the probability of each cell being occupied or vacant. The robot used this map, which was stored in its memory, to plan its trajectory and avoid obstacles. However, the default parameters of the navigation algorithms affected the robot's trajectory planning. It is possible to reduce trajectory drift and improve the efficiency of the robot's trajectory planning by adjusting the bias parameters for trajectory distance, goal distance, and occlusion scale.

Furthermore, this study also revealed the effectiveness of the navigation algorithms and planners implemented in the TurtleBot3 simulation environment. The configuration of the global planner, which includes the navfn and global planners, with a cost factor of 0.55, a neutral cost of 66, and a lethal cost of 253, proved to be highly desirable. This arrangement allowed for the generation of globally planned trajectories and effectively guided the robot to its target while avoiding environmental hazards. Local planners - including the DWA local planner, the eband local planner, and the teb local planner - also demonstrated their ability to produce the best velocity commands in light of the robot's immediate environment. After testing, it was discovered that the DWA local planner produced the most circular routes for the robot.

However, the effectiveness of trajectory planning algorithms is not always present in dynamic environments, and they are very difficult to generalize from a virtual to a real-world environment. The proposal to work with an MDP algorithm that falls within the domain of RL has led to very satisfactory results. After executing the script, the robot interacted with the virtual environment under the rules of Q-learning reinforcement learning. It will start with random actions and then begin to learn the best action to take in each state based on the rewards it receives. Training the robot through Q-learning takes time and requires a large amount of data for Turtlebot3 to learn effectively.

Moreover, the results of this study demonstrate the potential usefulness of SLAM and MDP algorithms for autonomous robots designed for hospital environments, particularly during the COVID-19 pandemic.

7 Conclusion

Throughout this article, we have explored the various techniques and algorithms used in robotics, with a particular focus on navigation, mapping, and localization. We have also seen how ROS and its various packages have made it easier to develop and test these algorithms, enabling us to perform simulations that closely mimic real-world scenarios. Moreover, we have examined the limitations of current techniques, highlighting the need for the development of algorithms that can enhance the autonomy of robots. In this regard, we discussed the potential benefits of combining SLAM algorithms with MDP to plan and execute robot trajectories. This study could be used to map and navigate hospital environments, deliver medical supplies and other essential items while reducing the risk of exposure for healthcare workers. These results highlight the potential of SLAM and MDP to play an important role in improving the safety and efficiency of healthcare delivery systems.

Overall, the field of robotics continues to evolve, and new techniques and algorithms are continually being developed to address the challenges posed by real-world applications. The use of ROS and its various packages has enabled researchers and developers to create highly sophisticated simulations that can test the efficacy of these algorithms in different environments. With further research and development, it is expected that robots will continue to play an increasingly significant role in various industries and fields, improving efficiency and productivity while reducing human labor.

Acknowledgment. The Project COVID-19 (2020–2022) has been funded with the support from the National Center for Scientific and Technical Research (CNRST) and Ministry of Higher Education, Morocco.

References

1. Drexler, N., Lapré, V.B.: For better or for worse: shaping the hospitality industry through robotics and artificial intelligence. Res. Hosp. Manage. **9**(2), 117–120 (2019)
2. Santosuosso, A., Bottalico, B.: Autonomous systems and the law: why intelligence matters. In: Robotics, Autonomics, and the Law, pp. 27–58. Nomos Verlagsgesellschaft mbH & Co. KG (2017)
3. Sadeghi Esfahlani, S., Sanaei, A., Ghorabian, M., Shirvani, H.: The deep convolutional neural network role in the autonomous navigation of mobile robots (SROBO). Remote Sens. **14**(14), 3324 (2022)
4. Vacariu, P.P.: Pirate robot autonomous navigation through complex pipe networks using reinforcement learning (2021)
5. Noori, F.M., Portugal, D., Rocha, R.P., Couceiro, M.S.: On 3D simulators for multi-robot systems in ROS: MORSE or Gazebo? In: 2017 IEEE International Symposium on Safety, Security and Rescue Robotics (SSRR), pp. 19–24. IEEE (2017)
6. Doroodgar, B., Liu, Y., Nejat, G.: A learning-based semi-autonomous controller for robotic exploration of unknown disaster scenes while searching for victims. IEEE Trans. Cybern. **44**(12), 2719–2732 (2014)
7. Bogue, R.: Sensors for robotic perception. Part two: positional and environmental awareness. Ind. Rob. Int. J. **42**(6), 502–507 (2015). https://doi.org/10.1108/IR-07-2015-0133
8. Karalekas, G., Vologiannidis, S., Kalomiros, J.: Europa: a case study for teaching sensors, data acquisition and robotics via a ROS-based educational robot. Sensors **20**(9), 2469 (2020)
9. Chikurtev, D.: Mobile robot simulation and navigation in ROS and Gazebo. In: 2020 International Conference Automatics and Informatics (ICAI), pp. 1–6. IEEE (2020)
10. Pajaziti, A.: Slam–map building and navigation via ROS. Int. J. Intell. Syst. Appl. Eng. **2**(4), 71–75 (2014)
11. Zhang, H., Zhang, C., Yang, W., Chen, C.Y.: Localization and navigation using QR code for mobile robot in indoor environment. In: 2015 IEEE International Conference on Robotics and Biomimetics (ROBIO), pp. 2501–2506. IEEE (2015)

12. Post, M.A., Bianco, A., Yan, X.T.: Autonomous navigation with open software platform for field robots. In: Gusikhin, O., Madani, K. (eds.) ICINCO 2017. LNEE, vol. 495, pp. 425–450. Springer, Cham (2020). https://doi.org/10.1007/978-3-030-11292-9_22

13. Ferreira, N.F., Araujo, A,, Couceiro, M.S., Portugal, D.: Intensive summer course in robotics–robotcraft. Appl. Comput. Informatics **16**(1/2), 155–179 (2018)

14. Govostes, R.Z., Littlefield, R.H., Jaffre, F., Kaeli, J.W.: Iterative software design for an autonomous underwater vehicle with novel propulsion capabilities and vision-based object tracking. In: OCEANS 2021: San Diego–Porto, pp. 1–4. IEEE (2021)

15. Quigley, M., Gerkey, B., Smart, W.D.: Programming robots with ROS: a practical introduction to the robot operating system. O'Reilly Media, Inc. (2015)

16. Hristozov, A.D., Matson, E.T., Gallagher, J.C., Rogers, M., Dietz, E.: Resilient architecture framework for robotic systems. In: 2022 International Conference Automatics and Informatics (ICAI), pp. 18–23 (2022)

17. Amsters, R., Slaets, P.: Turtlebot 3 as a robotics education platform. In: Merdan, M., Lepuschitz, W., Koppensteiner, G., Balogh, R., Obdržálek, D. (eds.) RiE 2019. AISC, vol. 1023, pp. 170–181. Springer, Cham (2020). https://doi.org/10.1007/978-3-030-26945-6_16

18. Cousins, S.: Willow garage retrospective [ros topics]. IEEE Robot. Autom. Mag. **21**(1), 16–20 (2015)

19. Xu, R., Li, C.: A review of high-throughput field phenotyping systems: focusing on ground robots. Plant Phenomics (2022)

20. Serrano-Muñoz, A., Elguea-Aguinaco, Í., Chrysostomou, D., BØgh, S., Arana-Arexolaleiba, N.: A scalable and unified multi-control framework for KUKA LBR iiwa collaborative robots. In: 2023 IEEE/SICE International Symposium on System Integration (SII), pp. 1–5. IEEE (2023)

21. Kirgis, F.P., Katsos, P., Kohlmaier, M.: Collaborative robotics. Robotic Fabrication in Architecture, Art and Design **2016**, 448–453 (2016)

22. Chung, S.Y., Huang, H.P.: Slammot-sp: simultaneous slammot and scene prediction. Adv. Robot. **24**(7), 979–1002 (2010)

23. Ravankar, A., Ravankar, A.A., Hoshino, Y., Emaru, T., Kobayashi, Y.: On a hopping-points svd and hough transform-based line detection algorithm for robot localization and mapping. Int. J. Adv. Rob. Syst. **13**(3), 98 (2016)

24. Zou, D., Tan, P., Yu, W.: Collaborative visual SLAM for multiple agents: a brief survey. Virtual Reality & Intelligent Hardware **1**(5), 461–482 (2019)

25. Thale, S.P., Prabhu, M.M., Thakur, P.V., Kadam, P.: ROS based SLAM implementation for autonomous navigation using turtlebot. In: ITM Web of Conferences, vol. 32, p. 01011. EDP Sciences (2020)

26. Kumar, S.: Development of SLAM algorithm for a Pipe Inspection Serpentine Robot (Master's thesis, University of Twente) (2019)

27. Looi, C.Z., Ng, D.W.K.: A study on the effect of parameters for ROS motion planner and navigation system for indoor robot. Int. J. Electr. Comput. Eng. Res. **1**(1), 29–36 (2021)

28. Ratliff, N., Zucker, M., Bagnell, J.A., Srinivasa, S.: CHOMP: gradient optimization techniques for efficient motion planning. In: 2009 IEEE International Conference on Robotics and Automation, pp. 489–494. IEEE (2009)

29. Zheng, K.: Ros navigation tuning guide. In: Koubaa, A. (ed.) Robot Operating System (ROS). SCI, vol. 962, pp. 197–226. Springer, Cham (2021). https://doi.org/10.1007/978-3-030-754 72-3_6

30. Chou, C.C., Lian, F.L., Wang, C.C.: Characterizing indoor environment for robot navigation using velocity space approach with region analysis and look-ahead verification. IEEE Trans. Instrum. Meas. **60**(2), 442–451 (2010)

Three-Phase Hybrid Evolutionary Algorithm for the Bi-Objective Travelling Salesman Problem

Omar Dib[✉][iD]

Wenzhou-Kean University, Wenzhou, China
odib@kean.edu

Abstract. In this research paper, we address the Bi-objective Traveling Salesman Problem (BTSP), which involves minimizing two conflicting objectives: travel time and monetary cost. To tackle this problem, we propose a novel three-Phase Hybrid Evolutionary Algorithm (3PHEA) that combines the Lin-Kernighan Heuristic, an enhanced Non-Dominated Sorting Genetic Algorithm, and a Pareto Variable Neighborhood Search. We conduct a comparative study comparing our approach with three existing methods specifically designed for solving BTSP. Our evaluation includes 14 instances of varying degrees of difficulty and different sizes. To assess the performance of the algorithms, we employ multi-objective performance indicators. The results of our study demonstrate that 3PHEA outperforms the existing approaches by a significant margin. It achieves coverage of up to 80% of the true Pareto fronts, indicating its superiority in solving the BTSP.

Keywords: bi-objective traveling salesman problem · hybrid evolutionary algorithms · non-dominated sorting memetic algorithm · pareto variable neighborhood search · multi-objective performance indicators

1 Introduction

The Multi-objective Traveling Salesman Problem (MTSP) has emerged as a prominent topic in combinatorial optimization, attracting considerable attention due to its wide range of applications across various domains [6]. For instance, in microchip manufacturing, MTSP is often used to minimize the travel time of robotic arms while simultaneously reducing production costs [3]. In the context of DNA sequencing, MTSP can be employed to efficiently plan the sequencing process, considering factors such as time constraints and the cost of sequencing different fragments [17]. In MTSP, the primary objective is to optimize multiple conflicting objectives simultaneously. These objectives vary depending on the specific problem context but often include travel time, distance, cost, energy consumption, or resource utilization. The goal is to find the best possible route that allows the traveler to visit a set of cities, each exactly once, and return to the starting point while minimizing the values of the chosen objectives [8].

R. El Ayachi et al. (Eds.): CBI 2023, LNBIP 484, pp. 173–186, 2023.
https://doi.org/10.1007/978-3-031-37872-0_13

The optimization of MTSP has garnered substantial interest from researchers and practitioners alike, as it presents a rich problem domain with numerous real-world applications. By addressing the challenges associated with MTSP, researchers aim to enhance decision-making processes in various fields, enabling more efficient resource allocation, cost reduction, and improved operational planning. In practice, solving the MTSP poses significant challenges due to the need to balance multiple objectives and the combinatorial nature of the problem. The task often involves searching through an exponentially ample solution space to identify the most desirable trade-off solutions that achieve a good compromise between the conflicting objectives. Thus, solving the MTSP requires the development of specialized algorithms and heuristics that can handle the complexity and provide high-quality solutions within a reasonable computational time [12].

Metaheuristic methods have garnered widespread interest in addressing the MTSP due to their ability to find high-quality solutions within reasonable computation times [7]. These approaches leverage iterative improvement strategies and heuristic search techniques to explore the vast solution space of the MTSP and identify near-optimal trade-off solutions. The use of metaheuristics offers several advantages in solving the MTSP. They balance solution quality and computational efficiency, enabling decision-makers to make informed decisions within practical time constraints. By combining exploration and exploitation strategies, metaheuristics effectively navigate the search space, continuously improving the solutions iteratively. Furthermore, the flexibility of metaheuristics allows for their adaptation to different problem characteristics and objectives [9].

Researchers have developed various metaheuristics, including Genetic Algorithms, Particle Swarm Optimization, Ant Colony Optimization, and Simulated Annealing, tailoring them to the specific requirements of the MTSP [5]. These approximate approaches are valuable tools for providing decision-makers with diverse, high-quality trade-off solutions. By harnessing the power of metaheuristics, researchers aim to overcome the computational infeasibility of exact algorithms and support efficient decision-making processes in various domains where the MTSP finds practical applications [13].

Combining metaheuristics has demonstrated excellent performance, as highlighted in a study by Dib et al. [10]. The authors proposed a novel approach that hybridized a Genetic Algorithm (GA) from the population-based metaheuristics domain with a variable neighborhood search (VNS). The authors conducted experiments using randomly generated and actual road network instances to evaluate their method's effectiveness. The experimental findings substantiated the efficiency of their approach when compared to the other methods considered.

Several approaches have been proposed to address the Multi-objective Traveling Salesman Problem (MTSP) and its variants, encompassing approximation and optimal techniques. For example, Lust and Teghem [16] introduced a two-phase Pareto local search (2PPLS) method for approximating the efficient set of the Bi-objective Traveling Salesman Problem (BTSP). However, their approach lacked evaluation against the actual Optimal Pareto fronts. In contrast, Florios and Mavrotas [11] presented AUGMECON2, an Augmented Epsilon Constraint Method, which was capable of generating the exact Pareto set for MTSP and other multi-objective integer programming problems. Although computationally intensive, AUGMECON2 successfully computed the actual Pareto fronts for various instances of the MTSP.

Population-based methods, including multi-objective evolutionary algorithms, have also been explored for MTSP [19]. Blank et al. [4] explored the Bi-Objective Traveling Thief Problem, which combines the Traveling Salesman Problem and Binary Knapsack Problem. They proposed different algorithms, including deterministic and evolutionary approaches. Their results offer insights into the properties of the bi-objective problems. Similarly, Moraes et al. [19] proposed a new evolutionary approach based on NSGA-II, SPEA2, and decomposition features for solving BTSP instances, demonstrating its effectiveness. However, their study did not provide details about the time complexity. Agrawal et al. [1] proposed a two-stage evolutionary algorithm (TSEA) for the Multi-objective Traveling Salesman Problem (MTSP), integrating a hybrid local search evolutionary algorithm. Their approach demonstrated superior performance to various genetic algorithm (GA) variants. However, the evaluation did not include a comparison with the actual Pareto fronts, which could introduce potential bias in the results.

In their work, Khan et al. [14] introduced a rule-based Artificial Bee Colony algorithm for solving the MTSP. Their approach determined the fitness function based on rules that followed the dominance property. To evaluate the performance of their method, the authors solved several instances from the TSP lib. They claimed that their approach was efficient for solving the MTSP. However, the assessment lacked a comparison with optimal algorithms, which weakened the evaluation. Additionally, the experimental estimate did not include a time indicator, thus limiting the analysis of the method's efficiency. Similarly, Michalak et al. [18] addressed the MTSP using an evolutionary algorithm incorporating random immigrants. These immigrants were introduced to enhance the diversity of the population and facilitate a more thorough exploration of the search space. The authors indicated and experimentally demonstrated that including random immigrants, combined with local search procedures within an evolutionary algorithm, resulted in a significant time overhead, limiting the approach's applicability. Moreover, they found that gradually decreasing the number of immigrants over time, akin to the cooling concept in simulated annealing, improved the results regarding both the hypervolume and the IGD indicators.

To address these gaps in the existing literature and provide a comprehensive analysis of efficient hybrid algorithms for MTSP, we introduce 3PHEA, a novel three-phase hybrid evolutionary approach. Our method leverages the Lin-Kernighan heuristic, known for its effectiveness in solving single-objective TSP, to generate an initial population. We then improve this population using a Hybrid Non-Dominated Sorting Genetic Algorithm (NSGA-II) and consolidate the solutions using a Pareto Variable Neighborhood Search. By developing 3PHEA and conducting an empirical analysis considering both time and quality perspectives, we aim to contribute to understanding MTSP and its efficient solution methodologies. Our approach combines the strengths of different metaheuristics to address the complexities of the BTSP and provide decision-makers with a diverse set of high-quality trade-off solutions.

2   Proposed Solution

The 3PHEA consists of three distinct phases to generate high-quality solutions. In the first phase, a set of supported non-dominated solutions is computed. This is followed by an improvement phase that utilizes a refined version of NSGAII. Finally, the third phase leverages the improved solutions using a Pareto Variable Neighborhood Search (VNS) strategy. Figure 1 provides a visual summary of the three steps, further elaborated in the subsequent subsections. These phases work harmoniously to tackle the bi-objective TSP and produce enhanced results.

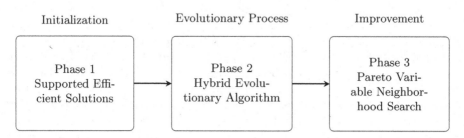

Fig. 1. Phases of the Three-Phase Hybrid Evolutionary Algorithm (3PHEA)

In Phase 1 of 3PHEA, an enhanced version of the Lust and Teghem [16] method is utilized. This phase aims to generate a set of supported efficient solutions by exploring all possible weight combinations. Each weight set is then used to perform a linear aggregation of objectives, solved using Lin-Kernighan-Helsgaun [21]. The implementation of this algorithm is integrated into Concorde, widely recognized as the most powerful solver for the TSP [2].

Algorithms 1 and 2 are followed to approximate the supported efficient solutions. The process begins by initializing the set $\hat{S}$ of supported efficient solutions with two initial solutions, namely lexmin $x \in X$ $(f1(x), f2(x))$ and lexmin $x \in X$ $(f2(x), f1(x))$. The BTSP is initially solved by considering only the first objective, assigning a weight vector of $(1, 0)$. The resulting solution is denoted as x_1. Similarly, another solution, x_2, is obtained using the weight vector $(0, 1)$.

Following the computation of x_1 and x_2, the dichotomic scheme outlined in Algorithm 2 is initiated. A normalization step is performed to handle the potentially large values of the weight sets, ensuring that $\lambda_1 + \lambda_2 = 1$. The coefficients of the matrix C^λ are rounded to the nearest integer value. The resulting solution, x_t, is added to $\hat{S}$ while only non-dominated solutions are retained in the set. This phase effectively leverages the modified Lust and Teghem method, integrating it with the Lin-Kernighan-Helsgaun algorithm implemented in Concorde. By utilizing these techniques, Phase 1 of 3PHEA generates a set of supported efficient solutions that form a strong foundation for subsequent phases of the algorithm.

Phase 2 of the 3PHEA algorithm presented in Algorithm 3 aims to approximate the set of non-supported efficient solutions using a hybrid evolutionary algorithm (HEA) that combines the non-dominated sorting genetic algorithm

---

Algorithm 1. Phase 1: Recursive Heuristic

Input: The cost matrices C^1 and C^2 of the bTSP;
Output: An approximation of the Pareto front $X_{\hat{SE1_m}}$;

1: *Solution* $x_1 \leftarrow$ solveTSP(C^1)   ▷*Apply heuristic algorithm to solve TSP on C^1*
2: *Solution* $x_2 \leftarrow$ solveTSP(C^2)   ▷*Apply heuristic algorithm to solve TSP on C^2*
3: $\hat{S} \leftarrow \{x_1, x_2\}$ ▷*Initialize set $\hat{S}$ with solutions x_1 and x_2*
4: $\hat{S} \leftarrow$ invokeRecursion($x_1, x_2, \hat{S}$)   ▷*Apply recursion-based method to improve solutions in $\hat{S}$*
5: $X_{\hat{SE1_m}} \leftarrow \hat{S}$ ▷*Set $X_{\hat{SE1_m}}$ as the final approximation of the Pareto front*

(NSGAII) and Pareto hill climbing (PHC). The HEA uses the approximated supported solutions obtained from Phase 1 as seeds to generate new supported and non-supported efficient solutions. The input parameters for HEA include an initial population P_0, a maximum number of generations G, the population size N, and the crossover and mutation rates c_r and m_r, respectively. Once P_0 is evaluated, the individuals are sorted based on their non-domination rank, followed by their crowding distance. Solutions with a lower rank, indicating proximity to the Pareto optimal front, are assigned a higher reproduction probability than those with a higher rank. The computational complexity of the non-dominated ranking procedure is $O(MN^2)$, where M is the number of objectives and N is the population size. Similarly, the crowding distance measure, used to preserve diversity, has a computational complexity of $O(MNlogN)$.

In the evolution process, a tournament selection mechanism is employed. Two solutions are compared based on their non-domination ranks, with the solution having a lower rank being preferred. If the two solutions have the same rank, the solution with the higher crowding distance is selected. The partially mapped (PMX) operator performs the crossover operation, although more advanced mechanisms can be applied, as suggested by Al et al. [15]. It is important to note that by following this strategy, the offspring solutions generated through crossover will always be feasible. Phase 2 of the 3PHEA algorithm effectively utilizes the HEA approach, integrating NSGAII and PHC, to approximate the non-supported efficient solutions. This phase combines the selection, crossover, and mutation operations to explore and diversify the population, generating a range of efficient solutions not supported by Phase 1.

After the crossover operation, Phase 2 of the 3PHEA algorithm utilizes a multi-objective version of hill climbing (HC-MO) [16] to generate new routes with improved quality and diverse variables. HC-MO employs a list of non-dominated solutions to guide the exploration of neighborhood solutions. These neighborhood solutions are identified using the fast non-dominated sort procedure of NSGAII. HC-MO can be visualized as a growing tree of non-dominated solutions. Initially, the tree is developed from different paths to uncover as many non-dominated solutions as possible from the list of neighbors. Subsequently, the tree is pruned to retain the best non-dominated solutions. Each path in the HC-MO tree has a root corresponding to the local search's initial solution. The path is then expanded

---

Algorithm 2. Phase 1: Solve Recursive Instance

Input: Two solutions x_1 and x_2;

Output: An approximation of the Pareto front $\hat{S}$;

1: $C^\lambda = \text{round}\left(\left[\lambda_1 c_{ij}^1 + \lambda_2 c_{ij}^2\right]\right)$ with

2: $\lambda_1 = \frac{f_2(x_r) - f_2(x_s)}{f_2(x_r) - f_2(x_s) + f_1(x_s) - f_1(x_r)}$ ▷*Compute λ_1 using objective values*

3: $\lambda_2 = 1 - \lambda_1$ ▷*Compute λ_2 as complement of λ_1*

4: *Solution* $x_t \leftarrow \text{solveTSP}(C^\lambda)$ ▷*Apply heuristic algorithm to solve TSP on C^λ*

5: $\hat{S} \leftarrow \text{addSolution}(\hat{S}, x_t)$ ▷*Add solution x_t to the set $\hat{S}$*

6: **if** $f_1(x_t) < f_1(x_s) \ \wedge \ f_2(x_t) < f_2(x_r)$ **then**   ▷*Check if x_t dominates x_s and is dominated by x_r*

7:      $\text{invokeRecursion}(x_r, x_t, \hat{S})$ ▷*Recursively solve with x_r and x_t*

8:      $\text{invokeRecursion}(x_t, x_s, \hat{S})$ ▷*Recursively solve with x_t and x_s*

9: **end if**

Algorithm 3. Phase 2: Workflow of Hybrid Evolutionary Algorithm

---

Input: $g := 1$ (Generation Counter);

$P_0 = $ (Initial Population);

$G = $ Maximum Numbre of Generations;

$N = $ Population Size;

$c_r = $ Crossover Rate;

$m_r = $ Mutation Rate;

Output: P_g

1: $P_g \leftarrow P_0$

2: **repeat**

3:      **evaluatePopulation**(P_g) ▷*Evaluate the fitness of each individual in P_g*

4:      **computeFastNonDominatedSort**(P_g) ▷*Perform fast non-dominated sorting on P_g*

5:      **computeCrowdingDistance**(P_g) ▷*Calculate crowding distances for individuals in P_g*

6:      $C_g \leftarrow \emptyset$

7:      **repeat**

8:          $p_1, p_2 \leftarrow \text{doSelection}(P_g)$ ▷*Select two parents from P_g*

9:          $c_1, c_2 \leftarrow \text{doCrossover}(p_1, p_2, c_r)$ ▷*Perform crossover on parents to generate offspring*

10:         $\widehat{C_1} \leftarrow \text{doMutation}(c_1, m_r)$ ▷*Apply mutation to offspring 1*

11:         $\widehat{C_2} \leftarrow \text{doMutation}(c_2, m_r)$ ▷*Apply mutation to offspring 2*

12:         $C_g \leftarrow C_g \cup \widehat{C_1} \cup \widehat{C_2}$ ▷*Add offspring to C_g*

13:      **until** $|C_g| \geq N$

14:      $C(g) \leftarrow P_g \cup C_g$

15:      **computeFastNonDominatedSort**(C_g) ▷*Perform fast non-dominated sorting on C_g*

16:      **computeCrowdingDistance**(C_g) ▷*Calculate crowding distances for individuals in C_g*

17:      $C_g \leftarrow \text{keepBest}(C_g, N)$ ▷*Keep the best N individuals from C_g*

18:      $P_{g+1} \leftarrow C_g$

19:      $g \leftarrow g + 1$

20: **until** $g \geq G$

21: **return** P_g

---

based on a predefined neighborhood structure. This expansion aims to explore the surrounding solutions and identify non-dominated solutions within the neighborhood. The offspring and parent populations are merged to determine the next generation's population, and a selection mechanism is applied to create the updated population C_g. The best N individuals, determined based on the domination rank and crowding distance rules, are selected from the parents and offspring populations to form the next generation's population. By employing HC-MO, Phase 2 of the 3PHEA algorithm effectively combines crossover with a multi-objective hill-climbing approach. This integration enables the generation of new solutions that are of better quality and exhibit diverse variable values. Using non-dominated solutions and the HC-MO tree structure contribute to exploring and identifying promising routes within the neighborhood, ultimately improving the population's quality for subsequent generations.

Algorithm 4. Three-Phase Hybrid Evolutionary Algorithm (3PHEA)

1: **procedure** PHASE1
2:     Generate weight combinations
3:     **for** each weight set **do**
4:         Perform linear aggregation
5:         Solve using Lin-Kernighan-Helsgaun algorithm
6:         Add supported efficient solutions to $\hat{S}$
7:     **end for**
8: **end procedure**
9: **procedure** PHASE2
10:     Perform hybrid evolutionary algorithm (HEA)
11:     Use supported solutions from Phase 1 as seeds
12:     Set input parameters for HEA
13:     Evaluate and sort individuals
14:     Select parents and perform crossover
15:     Mutate offspring solutions
16:     Update population C_g with offspring solutions
17:     Apply non-dominated sorting and crowding distance
18:     Keep the best solutions in C_g
19: **end procedure**
20: **procedure** PHASE3
21:     Apply Pareto Variable Neighborhood Search (PVNS)
22:     Initialize neighborhood index k to $k_{\min}$
23:     Create populations P_e, P, and P_a
24:     Explore neighbors and update P_e and P_a
25:     Update k based on non-dominated solutions found
26:     Repeat until all solutions of N_k are weakly dominated
27: **end procedure**

Phases 1 and 2 of the 3PHEA algorithm approximate the supported and non-supported efficient solutions. However, to fully exploit these solutions' diversity and high quality, we introduce a Pareto variable neighborhood search (PVNS) in Phase 3. The goal is to leverage the dynamic neighborhood feature of VNS to discover additional non-dominated solutions. In PVNS, we initialize the index k of the neighborhood structure with its minimum value, denoted as k_{min} (i.e., $k = k_{min}$). We maintain three populations: P_e to store efficient solutions, P as the current population, and P_a as an intermediate population used as an additional set. We add a variable for each solution s in P to indicate the neighbor functions executed on s. We then start the search for new non-dominated solutions by exploring all non-dominated solutions y for each solution x in P. If a neighbor solution y is not weakly dominated by the current solution x, it is added to the efficient solution population P_e. Similarly, if y is not weakly dominated by any solution $x \in P_e$, it is added to the intermediate population P_a.

After exploring all neighbor solutions of P concerning the current neighbor transformation N_k and adding them to P_a, we check if P_a is empty. If not, we set k back to k_{min}. However, if P_a contains solutions, non-dominated solutions can still be found based on the following neighbor structure. We increment k to move to the next neighbor structure in this case. This loop continues until all solutions from the neighbor structure N_k are weakly dominated by at least one solution in P_e. By performing this iterative process, PVNS aims to explore different neighborhoods and discover additional non-dominated solutions not captured in the previous phases.

This work applies four neighbor structures, specifically 1-*OPT*, 2-*OPT*, 3-*OPT*, and city insertion, in the PVNS phase. These neighborhood structures provide different ways to modify and explore the solutions to uncover new non-dominated solutions. By incorporating PVNS in Phase 3, the 3PHEA algorithm further improves the quality and diversity of the obtained solutions. The dynamic nature of the variable neighborhood search enables the algorithm to explore various neighborhoods and discover non-dominated solutions that were not previously identified, thereby enhancing the overall performance and effectiveness of the algorithm. A summary of 3PHEA is presented in Algorithm 4.

3   Experimental Study

To evaluate the performance of the proposed 3PHEA, a comprehensive analysis was conducted using 14 instances of the Multiple Traveling Salesman Problem (MTSP) from the work of Lust et al. [16]. For comparison purposes, two existing approaches were selected: AUGMECON2 (AUGM2), introduced by Florios et al. [11], and 2PPLS developed by Lust et al. [16]. These two approaches were chosen due to their comprehensive and detailed nature, making them the most prominent and well-documented methods in the literature. Moreover, the results obtained from AUGM2 and 2PPLS are openly available, which allows for a fair and transparent comparison. Several multi-objective performance indicators were employed to evaluate the performance of 3PHEA and compare it with AUGM2 and 2PPLS, as proposed by Riquelme et al. [20]. These indicators include Coverage, Hypervolume, and Epsilon, which comprehensively assess the algorithm's performance regarding solution quality and convergence.

Table 1. Experimental Settings

Parameter	Value
Initial Population	Concorde-Linkern
Population Size, and Generations	500
Crossover	PMX
Mutation	HC-MO
Crossover, and Mutation Probability	1
Selection	Tournament
HC-MO Iterations	10
HC-MO structures	2-OPT
PVNS k_{min}, and PVNS k_{max}	1, and 4
PVNS structures	$\{1, 2, 3\}$-OPT, City Insertion

Table 2. Empirical assessment for (Lust and Teghem [16]) datasets

Instance	Algorithm	$|\mathcal{PE}|$	$|\mathcal{D}|$	$|\mathcal{ND}| \uparrow$	$\mathcal{C} \uparrow$	$\mathcal{H} \uparrow$	$\mathcal{E} \downarrow$	$\mathcal{T} \downarrow$
L1	AUGM2	3332	0	3332	1	0.89944	0	134(h)
	2PPLS	2597	1043	1554	0.46638	0.89924	0.00157	25
	3PHEA	3112	474	2638	0.79171	0.89939	0.00104	271
L2	AUGM2	2458	0	2458	1	0.89463	0	74(h)
	2PPLS	1971	748	1223	0.49755	0.89449	0.00125	22
	3PHEA	2278	326	1952	0.79414	0.89460	$7.856e^{-4}$	182
L3	AUGM2	2351	0	2351	1	0.87799	0	49(h)
	2PPLS	1808	820	988	0.42024	0.87778	0.00150	20
	3PHEA	2190	499	1691	0.71926	0.87793	$9.070e^{-4}$	180
L4	AUGM2	2752	0	2752	1	0.88609	0	77(h)
	2PPLS	2163	792	1371	0.49818	0.88597	0.00101	24
	3PHEA	2598	363	2235	0.81213	0.88606	$5.519e^{-4}$	222
L5	AUGM2	2657	0	2657	1	0.87819	0	66(h)
	2PPLS	2014	610	1404	0.52841	0.87806	0.00111	25
	3PHEA	2435	313	2122	0.79864	0.87817	$6.347e^{-4}$	202
L6	AUGM2	2044	0	2044	1	0.892617	0	39(h)
	2PPLS	1661	604	1057	0.51712	0.89245	0.00139	21
	3PHEA	1869	262	1607	0.78620	0.89258	$7.580e^{-4}$	156
L7	AUGM2	1812	0	1812	1	0.86268	0	41(h)
	2PPLS	1361	540	821	0.45309	0.86244	0.00180	20
	3PHEA	1659	239	1420	0.78366	0.86263	$8.625e^{-4}$	141
L8	AUGM2	3036	0	3036	1	0.91943	0	52(h)
	2PPLS	2535	996	1539	0.50691	0.91928	0.00151	30
	3PHEA	2854	439	2415	0.79545	0.91940	$6.059e^{-4}$	241
L9	AUGM2	1707	0	1707	1	0.92987	0	34(h)
	2PPLS	591	318	273	0.15992	0.92934	0.00200	19
	3PHEA	982	465	517	0.30287	0.92966	0.00142	85
L10	AUGM2	1848	0	1848	1	0.88571	0	38(h)
	2PPLS	970	409	561	0.30357	0.88531	0.00213	24
	3PHEA	1378	332	1046	0.56601	0.88559	0.00152	110

Simulations were carried out on a computer system with the following specifications: the processor was an Intel(R) Core(TM) i7-10850H CPU with a clock speed of 2.70 GHz, and the system had 32.00 GB of installed RAM. The operating system was a 64-bit architecture. The experimental settings of the 3PHEA algorithm are summarized in Table 1. The simulations are repeated 10 times to ensure unbiased results, and the averages are taken. The parameter values are determined through non-extensive simulations. However, any tuning of the parameters is expected to primarily impact the convergence speed rather than the quality of the obtained approximations.

We summarize the 14 MTSP instances by indicating the ID, name, and number of cites of each instance, as follows: (L1, kroAB100), (L2, kroAC100), (L3, kroAD100), (L4, kroBC100), (L5, kroBD100), (L6, kroCD100), (L7, euclAB100), (L8, clusAB100), (L9, randAB100), (L10, mixdGG100), (F1, kroAB300), (F2, kroAB500), (F3, kroAB750), (F4, kroAB1000). Lust's datasets (L1-L10) have been used in [16], and Florios' datasets (F1-F4) have been used in [11].

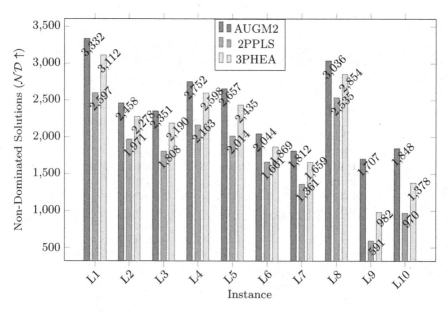

Fig. 2. Non-dominated solutions for AUGM2, 2PPLS, and 3PHEA algorithms

Table 2 presents the results of 3PHEA compared to AUGMECON2 (AUGM2) ([11]) and 2PPLS ([16]) for instances (L1 to L10). $|\mathcal{PE}|$ represents the number of potentially efficient solutions, $|\mathcal{D}|$ number of dominated solutions, $|\mathcal{ND}|$ number of non-dominated solutions, $\mathcal{C}$ Coverage metric, $\mathcal{H}$ Hypervolume, $\mathcal{E}$ Epsilon indicator, and $\mathcal{T}$ time in seconds unless specified. AUGM2 is an exact approach for multi-objective problems based on an improved epsilon constraint and advanced branch and cut techniques. On the other hand, 2PPLS is based on a Pareto local search procedure dedicated to MTSP and results in an approximated Pareto front. Hence the optimal results of AUGM2 are used as a reference.

As seen from Table 2, the computational time of AUGM2 is very high, ranging from 34 to 134 h, depending on the instance. AUGM2 execution time is often exponentially proportional to $|\mathcal{PE}|$. Hence, although the instance size is small, AUGM2 might drastically suffer from high computation time for instances with many non-dominated solutions. As for 2PPLS, the computation time is relatively low, ranging between 20 and 30 s, and the coverage metric varies between 0.15 and 0.52. For example, for the instance L9 that is randomly generated, 2PPLS only found 273 among 1707 solutions; the coverage improved but was still relatively low for L10, that is has a mixture of random and Euclidean costs; lastly, for the Euclidean instances, 2PPLS coverage relatively improves to reach 0.52 such as in instance L5. Those results indicate that the local search procedure in 2PPLS might rapidly fall into local minima as the instance is random. As far as 3PHEA is concerned, the computation time follows a similar trend compared to AUGM2 and 2PPLS. Assessing the coverage metric indicates that the proposed 3PHEA method could significantly approximate the true Pareto front compared to 2PPLS. Specifically, 3PHEA achieves up to 81% of the non-dominated solutions compared to 49% resulting from 2PPLS (see instance L4 in Table 2). Therefore, 3PHEA-obtained fronts are significantly superior and closer to the actual Pareto fronts. The number of non-dominated solutions ($\mathcal{ND}\uparrow$) for all algorithms and instances is visualized in Fig. 2.

Table 3. Empirical assessment for (Florios and Mavrotas [11]) datasets

Instance	Algorithm	$|\mathcal{PE}|$	$|\mathcal{D}|$	$|\mathcal{ND}|\uparrow$	$\mathcal{C}\uparrow$	$\mathcal{H}\uparrow$	$\mathcal{E}\downarrow$	$\mathcal{T}\downarrow$
F1	AUGM2	14867	0	14867	1	0.91922	0	-
	2PPLS	15092	4797	10295	0.69247	0.91863	$9.361e^{-4}$	205
	3PHEA	16637	428	16209	1.09026	0.91868	$7.080e^{-4}$	1497
F2	AUGM2	33929	0	33929	1	0.92829	0	-
	2PPLS	1932	707	1225	0.48418	0.86352	0.00127	458
	3PHEA	38109	3143	34966	1.03056	0.92838	$8.460e^{-5}$	7164
F3	AUGM2	61184	0	61184	1	0.93790	0	-
	2PPLS	60657	20831	39826	0.65092	0.92986	$6.345e^{-5}$	3256
	3PHEA	67301	9488	57813	0.94490	0.93687	$4.411e^{-5}$	8803
F4	AUGM2	98151	0	98151	1	0.94433	0	-
	2PPLS	45120	0	45120	0.45969	0.91174	$6.345e^{-5}$	7462
	3PHEA	95128	7872	87256	0.88899	0.93543	$4.411e^{-5}$	12604

Results of larger instances F1, F2, F3, and F4 from Florios and Mavrotas [11], corresponding to 300, 500, 750, and 1000 cities, respectively, are presented in Table 3, and Fig. 3. It is important to note that the true Pareto fronts for these instances are not yet known in the literature. Therefore, the performance of the algorithms is evaluated based on various multi-objective indicators.

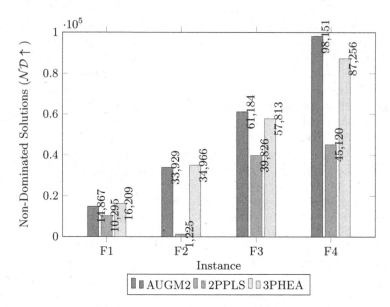

Fig. 3. Comparison of ($\mathcal{ND}$) Solutions for Different Algorithms and Instances

Table 3 and Fig. 3 present the outcomes of three algorithms: AUGM2, 2PPLS, and 3PHEA. The table provides information on the number of solutions in the Pareto front ($|\mathcal{PE}|$), dominated solutions ($|\mathcal{D}|$), non-dominated solutions found ($|\mathcal{ND}|\uparrow$), coverage ($\mathcal{C}\uparrow$), hypervolume ($\mathcal{H}\uparrow$), average epsilon ($\mathcal{E}\downarrow$), and average computation time ($\mathcal{T}\downarrow$) for each instance. Notably, AUGM2, serving as a reference method for 2PPLS and 3PHEA, does not yield the true Pareto front due to its high computation time. Instead, AUGM2 utilizes an approximation approach based on Pareto local search and simulated annealing. The results reveal that 3PHEA surpasses AUGM2 in terms of coverage, indicating its ability to discover novel non-dominated solutions that AUGM2 misses. For instance F1, 3PHEA identifies 16209 non-dominated solutions, whereas AUGM2 and 2PPLS find 14867 and 10295 solutions, respectively. This demonstrates the superior performance of 3PHEA in uncovering a greater number of non-dominated solutions. However, for instances F3 and F4, the coverage of 3PHEA is slightly lower than AUGM2 due to the absence of certain optimization techniques. Nonetheless, 3PHEA outperforms 2PPLS across various multi-objective indicators. The results also underscore the influence of instance size on computation time, with larger instances, like F4 with 1000 cities, resulting in lengthier execution times for all methods. For instance, the execution time of 3PHEA for F4 increases to 12604 s. This duration would be even greater if more neighborhood structures were considered.

4   Conclusions

In conclusion, this paper presented a novel three-Phase Hybrid Evolutionary Algorithm (3PHEA) for solving the Bi-objective Traveling Salesman Problem (BTSP). The algorithm was evaluated on 14 BTSP instances, and its performance was analyzed using various performance indicators. The results demonstrated the significant superiority of 3PHEA compared to existing approaches, as it was able to cover up to 80% of the actual Pareto fronts. In future research, several enhancements and extensions are planned. Firstly, additional criteria such as vehicle capacity and the uncertain travel time will be incorporated into the algorithm to address more realistic and complex scenarios. This will enhance the algorithm's applicability in practical transportation optimization problems. Furthermore, efforts will be made to optimize the time complexity of the neighborhood structures used in the algorithm. This optimization will be achieved by incorporating a machine-learning module, enabling the algorithm to learn and adapt to problem-specific characteristics and improve efficiency. By considering these future directions, the proposed 3PHEA algorithm holds great promise for advancing the field of BTSP and contributing to more efficient and effective solutions for real-world transportation optimization challenges.

References

1. Agrawal, A., Ghune, N., Prakash, S., Ramteke, M.: Evolutionary algorithm hybridized with local search and intelligent seeding for solving MTSP. Expert Syst. Appl. **181**, 115192 (2021)
2. Applegate, D., Bixby, R., Chvatal, V., Cook, W.: Concorde TSP solver (2006). http://www.tsp.gatech.edu/concorde
3. Bergel, A., Bergel, A.: The traveling salesman problem. Agile Artificial Intelligence in Pharo: Implementing Neural Networks, Genetic Algorithms, and Neuroevolution, pp. 209–224 (2020)
4. Blank, J., Deb, K., Mostaghim, S.: Solving the bi-objective traveling thief problem with multi-objective evolutionary algorithms. In: Trautmann, H., et al. (eds.) Evolutionary Multi-Criterion Optimization, pp. 46–60 (2017)
5. Cai, X., Wang, K., Mei, Y., Li, Z., Zhao, J., Zhang, Q.: Decomposition-based Lin-Kernighan heuristic with neighborhood structure transfer for multi/many-objective traveling salesman problem. IEEE Transactions on Evolutionary Computation, pp. 1 (2022). https://doi.org/10.1109/TEVC.2022.3215174
6. Cheikhrouhou, O., Khoufi, I.: A comprehensive survey on the multiple traveling salesman problem: applications, approaches and taxonomy. Comput. Sci. Rev. **40**, 100369 (2021). https://doi.org/10.1016/j.cosrev.2021.100369
7. Dib, O., Moalic, L., Manier, M.A., Caminada, A.: An advanced GA-VNS combination for multicriteria route planning in public transit networks. Expert Syst. Appl. **72**, 67–82 (2017). https://doi.org/10.1016/j.eswa.2016.12.009
8. Dib, O.: Novel hybrid evolutionary algorithm for bi-objective optimization problems. Sci. Rep. **13**(1), 4267 (2023). https://doi.org/10.1038/s41598-023-31123-8
9. Dib, O., Caminada, A., Manier, M.A., Moalic, L.: A memetic algorithm for computing multicriteria shortest paths in stochastic multimodal networks. In: Proceedings of the Genetic and Evolutionary Computation Conference Companion, pp. 103–104 (2017)

10. Dib, O., Manier, M.A., Moalic, L., Caminada, A.: Combining VNS with genetic algorithm to solve the one-to-one routing issue in road networks. Comput. Oper. Res. **78**, 420–430 (2017). https://doi.org/10.1016/j.cor.2015.11.010
11. Florios, K., Mavrotas, G.: Generation of the exact pareto set in MTSP and set covering problems. Appl. Math. Comput. **237**, 1–19 (2014)
12. George, T., Amudha, T.: Genetic algorithm based multi-objective optimization framework to solve traveling salesman problem. In: Sharma, H., Govindan, K., Poonia, R.C., Kumar, S., El-Medany, W.M. (eds.) Advances in Computing and Intelligent Systems. AIS, pp. 141–151. Springer, Singapore (2020). https://doi.org/10.1007/978-981-15-0222-4_12
13. Jin, Z., Dib, O., Luo, Y., Hu, B.: A non-dominated sorting memetic algorithm for the multi-objective travelling salesman problem. In: 2021 4th International Conference on Algorithms, Computing and Artificial Intelligence, pp. 1–6 (2021)
14. Khan, I., Maiti, M.K., Basuli, K.: MTSP: an ABC approach. Appl. Intell. **50**(11), 3942–3960 (2020)
15. Kumar, R.: A survey on memetic algorithm and machine learning approach to traveling salesman problem. Int. J. Emerg. Technol. **11**(1), 500–503 (2020)
16. Lust, T., Teghem, J.: Two-phase pareto local search for the BTSP. J. Heuristics **16**(3), 475–510 (2010)
17. Mandal, A.K., Kumar Deva Sarma, P.: Novel applications of ant colony optimization with the traveling salesman problem in DNA sequence optimization. In: 2022 IEEE 2nd International Symposium on Sustainable Energy, Signal Processing and Cyber Security (iSSSC), pp. 1–6 (2022). https://doi.org/10.1109/iSSSC56467.2022.10051206
18. Michalak, K.: Evolutionary algorithm using random immigrants for the MTSP. Procedia Comput. Sci. **192**, 1461–1470 (2021)
19. Moraes, D.H., Sanches, D.S., da Silva Rocha, J., Garbelini, J.M.C., Castoldi, M.F.: A novel multi-objective evolutionary algorithm based on subpopulations for the BTSP. Soft Comput. **23**(15), 6157–6168 (2019)
20. Riquelme, N., Von Lücken, C., Baran, B.: Performance metrics in multi-objective optimization. In: 2015 Latin American Computing Conference, pp. 1–11 (2015)
21. Zheng, J., He, K., Zhou, J., Jin, Y., Li, C.M.: Combining reinforcement learning with Lin-Kernighan-Helsgaun algorithm for the traveling salesman problem. In: Proceedings of the AAAI Conference on Artificial Intelligence, vol. 35, pp. 12445–12452 (2021)

Understanding Data Journalism Acceptance Among Social TV Users: A Case Study of Twitter in the United Arab Emirates

Faycal Farhi[1]([⊠]) [iD], Riadh Jeljeli[1], Abdullah Ali Al Marei[2], and Khaled Zamoum[3]

[1] College of Communication and Media, Al Ain University, Al Ain, United Arab Emirates
{faycal.farhi,Riadh.jeljeli}@aau.ac.ae

[2] Media and Communications Department, Faculty of Humanities, King Khalid University, Abha, Saudi Arabia
aalmarai@kku.edu.sa

[3] College of Communication, University of Sharjah, Sharjah, United Arab Emirates
Kzamoum@sharjah.ac.ae

Abstract. Today Social TV users prefer data journalism for different reasons. One of the basic reasons behind this behaviour is audiences' growing interest in transparent and detailed news reports. This research also examined the factors proposed by the Uses and Gratifications theory and the Technology Acceptance Model in the acceptance of data journalism among Emirati Social TV users. Data gathered from structured survey questionnaires were analyzed by using Structural Equation Modelling. Results showed that Information Sharing and Information Seeking are two significant factors linked to users' gratifications. These gratifications, on the other hand, significantly affect their data journalism acceptance. Besides, the mediating effect of perceived usefulness also remained significant, indicating the expected useful outcomes are important determinants of their data journalism acceptance. UOG and TAM significantly predict data journalism acceptance among Emirati Social TV audiences. Thus, it is concluded that data journalism has become a valuable source of information for social TV users, providing detailed insights into complicated issues in an easily understandable and engaging format. Besides, perceived ease of use remains an important factor in social TV users' acceptance of data journalism, as it contributes to their willingness to engage with this content. Thus, as the social TV audience evolves from Twitter users to other platforms, it is usefully integrated with data journalism.

Keywords: Data Journalism · Social TV · Twitter · United Arab Emirates · Uses and Gratifications · Technology Acceptance Model

1 Introduction

Data journalism is crucial to modern journalism, particularly for online media users [1]. The internet has given rise to abundant information, and data journalism is vital in providing precise and reliable information to online media users [2]. According to Zhang & Chen [3], data journalism provides in-depth analysis that classic reporting methods may not offer. Today, data journalism helps online media users understand complex issues that would otherwise be difficult to comprehend by analyzing data and presenting it easily and visually compellingly. Camaj et al. [4] consider data journalism as promoting transparency and accountability in journalism. Data journalists allow readers to verify the information presented and build trust with their audience by providing transparency. This transparency in data journalism encourages accuracy, objectivity, and accountability, making it an essential practice for contemporary journalism [5]. According to Bebawi [6], data journalism comprises evidence-based journalism that cannot be easily overlooked to support or reveal stories. As online media users can entrust that the information, they receive from data journalism is based on facts, data journalism makes it easier to hold those in power accountable. Especially in a world where misinformation and fake news have become rampant, data journalism can be of greater significance [7]. Besides, data journalism also provides engaging and interactive content that online media users find attractive. Interactive maps, charts, and infographics used in data journalism allow online media users to explore data in a visually appealing and interactive way. As a result, data journalism becomes interesting and enhances the understanding and retention of information [8].

However, as technology evolves [9], the conventional patterns of news watching, gathering, and analyzing information are also progressing [10]. Talking specifically about data journalism for online media users, many platforms are encouraging exposure to data journalism through different perspectives. For instance, Twitter is much used for data journalism because of its real-time nature and the vast amount of user-generated data. Data journalists uncover insights into the audience's reactions and opinions by analyzing tweets about different TV shows [11]. This data can be used to identify trending topics, track the popularity of characters or storylines, and even anticipate the results of reality TV shows. Talking specifically about Twitter usage by Social TV users, they rely much on data journalism especially Twitter users. Social TV refers to data on social media platforms to discuss and enrich the venture of watching television programs. Twitter is one of the most widely used platforms among social TV users. By collecting and analyzing data generated by social TV users, data journalism can obtain valuable insights and analyze various aspects of television programs [12]. As a result, today, social TV frequently relies on Twitter for news-gathering purposes. They further share and re-share tweets, retweets, likes, and replies based on the information they receive from data journalism-based reports. According to Widholm & Appelgren [12], data journalism also helps social TV users to improve their critical thinking abilities as data journalists identify emerging trends and social issues that may have remained hidden or underrepresented due to certain circumstances.

Study Gaps and Aims

Although data journalism is not new, its adoption and usage surged during the Covid-19 pandemic (Auväärt, 2022; Desai et al., 2021). Consequently, it became strongly considered, especially by practitioners and media organizations. However, the relevant approach is still being used today, further demanding a need to analyze its constant usage during the post-pandemic era, indicating a gap in the current literature. Thus, this research examines the causal factors behind data journalism acceptance by Social TV users under the Uses and Gratifications theory and Technology Acceptance Model.

The focus would remain on data journalists in the United Arab Emirates to provide region-specific results, which will be represented systematically and empirically.

2 Related Literature

2.1 Data Journalism in the United Arab Emirates - A Brief Review

Arab journalism has faced much criticism due to the presence of authoritarian governments, leading to self and state-imposed censorship, making it challenging for journalists to report freely [15]. Many journalists have started using data-driven journalism to combat this, offering more reliable, efficient, and accessible practices. Remarkably, data journalism is widely accepted by Emirati journalists who believe in the normative practice of press freedom. Today, over 85% of Emirati journalists practice data journalism, considering it a sought-after professional practice [16]. The use of data journalism in the United Arab Emirates is essential for two reasons [17]. First, published literature suggests that data journalism is highly understandable for readers. Second, data journalism empowers Emirati journalists to address different socio-political issues without constraints. Emirati journalists use code-free browser extensions or programming languages such as Python to access locked data or acquire data from the application programming interface. As a result, Emirati journalists prioritize freedom of expression and access to information as fundamental human rights, highlighting the significance of data journalism for them [18].

2.2 Information Sharing and Gratifications

According to Appelgren et al. [19], one of the immediate goals of news and data journalism is to provide factual news reports to the public seeking suitable resources for information seeking. Journalists use a variety of sources, including interviews, documents, and data, to verify their stories' accuracy. They also stick to ethical standards, such as protecting sources and avoiding conflicts of interest. De-Lima-Santos & Mesquita [20] argued that data journalists have immediate access to information gathering and spreading information, indicating an important aspect of today's adoption of data journalism. For example, data journalists can instantly gather sensitive news, and their role is to hold those in power accountable. By providing information about government actions, corporate behaviour, and other essential issues, journalists help ensure social stability.

H1: Information sharing has a positive effect on gratification among Emirati audiences.

2.3 Information Seeking and Gratifications

Data journalism involves using data analysis techniques to uncover and report stories. To adopt data journalism, it's crucial to have a solid understanding of the data sources, data analysis approaches, coding, data visualization tools, storytelling, and others [6]. According to Arias-Robles & López [21], the prerequisites of data journalism make it more compelling as they ensure access and exposure to important datasets. As a result, data journalists provide the audience with information that is more interesting, appealing, and based on factual details [22]. As noted by Allen & Pollack [23], data journalism allows information sharing through data-driven stories that ensure the availability of supplementary information in reporting. This approach allows journalists to extract insights from large datasets, revealing trends and patterns that might go unnoticed.

H2: Information seeking has a positive effect on gratification among Emirati audiences.

2.4 Convenience and Gratifications

According to Desai et al. [14], data journalism is considered a convenient approach for journalists due to different factors that lead to its wider adoption. Geng et al. [24] stated that data journalism is becoming more readily available, with many organizations publishing online datasets in open data sources. Mitchell et al. [25] further noted data journalism tools and techniques have become more accessible, even for those without specialized technical skills, leading to representing more engaging stories with interactive elements and attracting a larger audience. According to Erduran & Demirkiran [22], data journalism permits a more systematic and stringent reporting approach, leading to more accurate and impactful journalism. This can be particularly important in a world where trust in the media is increasingly under threat. Today, audiences accept data journalism because it provides a more personalized experience using data to create interactive tools and visualizations. Data journalism allows audiences to explore issues tailored to their interests and needs, leading to a more engaged and satisfied audience, who are more likely to return to the same news source [26].

H3: Convenience has a positive effect on gratification among Emirati audiences.

2.5 Gratifications and Acceptance Behavior

Data journalism has achieved significant traction in recent years. It involves using simple yet comprehensive data analysis approaches to uncover trends, patterns, and insights that might take time to be apparent from raw data [27]. Data journalists have command over working with large data sets, which adds value to adopting it as a modern journalistic approach. According to Showkat & Baumer [28], audiences accept data journalism as it provides a more comprehensive understanding of complex issues, enabling audiences to gain a deeper insight into the world around them. Data journalists engage audiences more meaningfully than traditional journalism by presenting data-driven stories in an accessible and visually appealing way. As a result, the audience appreciates data journalists, that further increases a deeper understanding of the value of quality journalism. According to (Tandoc et al. [29], data journalism also provides a more objective and fact-based approach to reporting, helping to counter the spread of misinformation and fake news. Hofmann et al. [30] argued that one of the primary reasons behind broad data journalism acceptance is that data journalists help build trust between journalists and their audiences.

H4: Gratifications have a positive effect on the acceptance behaviour of Emirati audiences.

2.6 Perceived Usefulness in Data Journalism Acceptance

According to Mutsvairo [31], data journalism provides a more efficient and cost-effective way of accessing news reports. By leveraging data to uncover insights and trends, data journalism enables audiences to focus and read about the issues that might otherwise be too time-consuming or challenging to access in traditional ways. As a result, the audience experiences democratized news as it is accompanied by more accessible and affordable to a broader range of people. Figueiras et al. [32] argued that data journalism provides audiences with an engaging and interactive experience. Data journalism allows audiences to explore issues more personalized and interactively by using data to create visualizations, charts, and interactive tools, leading to a more satisfying and engaging experience for users likelier to return to the same news source in the future.

H5: Perceived usefulness moderates the relationship between the acceptance behaviour of Emirati audiences.

3 Theoretical Background

This research is based on the uses and gratification theory that magnifies the role of the audience in selecting certain media and their content to meet their requirements [33, 34, 35]. Besides, a latent construct from Technology Acceptance Model [36, 37] is also incorporated to provide a comprehensive picture behind data journalism acceptance among Twitter users in the United Arab Emirates. Journalism caters to audiences' information gathering and seeking needs in several ways by providing audiences with timely and relevant information about current events and issues. Information seeking and gathering are two critical needs behind specific media selections affirmed by existing literature. As a result, the audience can stay informed and engaged with the world around them. Talking about data journalism helps audiences understand complex issues and topics by breaking them down into more digestible and understandable pieces. By using clear and accessible language, journalism can make it easier for audiences to comprehend complex issues and make informed decisions. [38] Also, data journalism acceptance as accompanied by its capability to uncover issues attributed to marginalized communities and gain less consideration. Consequently, the audience may consider it a modern, objective, and comparatively better journalistic approach that also fulfils the conceptualization of journalism as the "fourth pillar of state" (Perceived usefulness). Fahmy & Attia [16] further argued that data journalists provide audiences with a sense of community and shared experience. By reporting on issues that affect people's lives and communities, journalism can help to bring people together and foster a sense of shared understanding and purpose. Thus, based on the cited literature and theoretical support, Fig. 1 illustrates the explanatory model of current research. The role of social TV users, particularly on Twitter, is believed to be vital in data journalism. Data journalists utilize reports to provide online users with insights and analysis of television programs, detect trends, monitor audience reactions and opinions, and reveal emerging storylines [8]. This highlights the importance of social TV users as a trending type of audience and their association with data journalism.

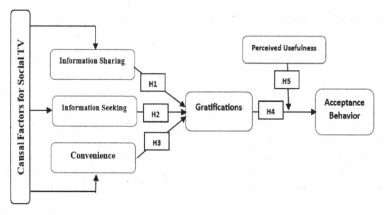

Fig. 1. Explanatory Framework of Current Research

4 Methods

The research design used in this study is a case study based on obtaining data from individuals from a certain setting (Twitter Usage and Social TV exposure). The researchers applied structured questionnaires as the aim was to obtain data ensuring generalizability and valuable insights regarding the phenomenon under study [39]. The survey was based on items and scales adopted from some existing studies. Notably, the adopted content was first edited to match the study objectives and further checked for reliability and validity as a part of Structural Equation Modelling [40]. Table 1 indicates the sources of measures and scales adopted for the current research. The relevant table also indicates the reliability analysis indicating all the Cronbach Alpha and Composite Reliability values surpassing .7 as the minimum threshold value. The survey was conducted using a Five-Point Likert scale questionnaire [39, 41] from July to August 2022. The data was calculated using SPSS and Amos for Structural Equation Modelling (SEM) [42].

Table 1. Sources of Survey Measures and Items

S/R	Items and Scales	Source	CA	CR
1	Information Sharing	[43]	.738	.837
2	Information Seeking	[44]	.831	.820
3	Convenience	[45]	.790	.713
4	Perceived Usefulness	[46]	.719	.785 ·
5	Gratifications	[47]	.801	.740
6	Acceptance Behavior	[41]	.793	.800

4.1 Study Populations and Sampling Method

The population for this study consisted of Twitter users in the United Arab Emirates. Recent data indicates that currently, there are 4.56 million Twitter users in the UAE, with daily 3.9 million users countrywide [48]. Thus, based on the available data, a sample of $n = 384$ users was selected using Krejci and Morgan's sample selection formula [49]. However, the researchers selected only those Twitter-based Social TV users exposed to data journalism. Despite its limitations, the convenience sampling approach was the most suitable to match the study objectives and problem. The respondents were provided with informed consent, and data privacy was also ensured as one of the core research ethics. After data collection, the responses were carefully evaluated for the final analysis. The researchers found 14 questionnaires missing/ not sent back by the respondents, indicating a total response rate of 96.3%, greater than the minimum response rate of 60% [50].

5 Analysis and Findings

The study analysis is based on both inferential and descriptive statistics. The gathered data was carefully coded and analyzed. Table 2 summarizes the results of the descriptive analysis. Further, the inferential statistics were based on Structural Equation Modelling, including inner and outer model analyses.

Table 2. Descriptives of Gathered Data

Constructs	Mean	SD	Min	Max	Bootstrapping 95% Confidence Interval	
					Lower	Upper
Sharing	3.91	.239	2.00	.400	370	370
Seeking	3.85	.290	3.00	4.00	370	370
Convenience	3.91	.200	2.67	4.00	370	370
Gratifications	3.92	.206	3.00	4.00	370	370
Usefulness	3.39	.462	3.00	4.00	370	370
Acceptance	3.87	.287	2.67	4.00	370	370

Multicollinearity is an important factor, considered probable in studies having more than one predictor variable. According to Asthana [51], multicollinearity is one of the most unlikeable phenomena in multiple regression-based studies. Thus, the relevant analysis was also conducted in the current research using the Variance Inflation Factor analysis. Results showed the VIF value for Information Sharing was 1.535, Information Seeking 1.183, Convenience 1.238, and Gratifications was 1.306. All the obtained values were lower than the minimum cutoff value of 3.0, indicating that multicollinearity between the predictors is under control (See Table 3).

The inner model analysis was conducted to examine the internal consistency between the constructs, also known as convergent validity. First, results showed that all the factors

Table 3. Analysis of Multicollinearity between Predictor Variables

Constructs	TOL	VIF	Sign
Sharing	.652	1.535	.001
Seeking	.846	1.183	.390
Convenience	.808	1.238	.001
Gratifications	.765	1.306	.000

loads are greater than the threshold value of .5. Besides, calculating the average of each construct revealed the Average Variance Extracted values are also greater than the threshold value of .5. Overall, the internal consistency between the constructs is affirmed (See Table 4).

Table 4. Analysis of Convergent Validity

Constructs	Items	Loads	AVE
Information Sharing	INS1	.707	.753
	INS2	.817	
	INS3	.735	
Information Seeking	INE1	.751	.691
	INE2	.656	
	INEE	.666	
Convenience	CON1	.783	.677
	CON2	.617	
	CON3	.631	
Gratifications	GRA1	.638	.640
	GRA2	.511	
	GRA3	.643	
Perceived Usefulness	USL1	.688	.739
	USL2	.791	
	USL3	.515	
Acceptance Behavior	BEH1	.796	.695
	BEH2	.595	
	BEH3	.582	

The table shows the person correlation coefficient values to compare them with the square of AVE values. Notably, the relevant analysis determined the extent to which study constructs correlate with them [52]. Analysis showed that all the squares of AVE

values are greater than the correlation values and are distinct from each other. Overall,
the discriminant validity is also affirmed (See Table 5).

Table 5. Pearson Correlation Coefficient

	INS	INE	CON	GRA	USF	BEH
INS	**.567**					
INE	.326	**.477**				
CON	.387	.311	**.450**			
GRA	.467	.084	.216	**.409**		
USL	−.048	−.039	−.064	−.111	**.586**	
BEH	.060	.064	.191	.308	.116	**.483**

The goodness of fit was analyzed to examine the extent to which sample data fits
well with the expected data [53]. The relevant analysis was based on excluding one of
the items (GRA2) having a lower factor loading value to nullify its potential impacts on
the structural model analysis. Results showed the Standardized Mean Square value of
.153, lower than the threshold value of 0.85 [40]. Besides, the Tucker and Lewis value
was .973, and the probability level was at .000. Overall, the results showed a good fit
for the study model. Figure 2 shows the measurement model used for the goodness of
fit analysis.

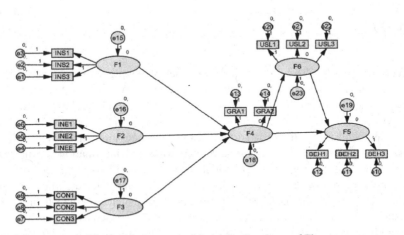

Fig. 2. Measurement Model for Goodness of Fit

The outer model analysis was conducted to examine the proposed hypotheses. The
relevant analysis was based on applying the path analysis accompanied by regression
weights [54]. Table 6 shows the results of the path analysis. First, the hypothesis assumed
a direct effect of Information Sharing on the Gratifications and was found to be validated

with the significance value of p <.003. The second hypothesis proposed a significant effect of Information Seeking on the Gratifications, which also remained approved by the analysis (p <.000). However, the third hypothesis was rejected as the effect of Convenience on Gratifications remained insignificant with the p-value of p >.838. The fourth study hypothesis assuming the effect of Gratifications, also remained significant as the p-value was at .001, lower than the threshold value of 0.05. Finally, the mediation of Perceived Useful on the relationship between Gratifications and Acceptance Behavior was examined. Notably, the effect of Information sharing on gratifications was already validated. Thus, the mediating effect of Perceived Usefulness also remained significant with the p-value at p <.000). Furthermore, path analysis showed the strongest path between Information Seeking and Gratifications (1.360), followed by Perceived Usefulness and Acceptance Behavior (.639), the Information Sharing and Gratification had the third most robust path value. Overall, the results remained supportive except for the H3 of the study.

Table 6. Hypothesis Testing and Path Analysis

Hypothesis	β	Sign	T	Decision
Information Sharing → Gratifications	.523	.003	4.728	Accepted
Information Seeking → Gratifications	1.360	.000	13.752	Accepted
Convenience → Gratifications	.036	.928	.838	Rejected
Gratifications → Acceptance Behavior				Accepted
Hypothesis	β	Sign	Indirect Effects	Decision
Gratifications → Perceived Usefulness → Acceptance Behavior	.639	.002	.736	Accepted

According to the Uses and Gratifications theory, people actively consume media based on their needs and gratifications. Besides, the Technology Acceptance Model (TAM) helps explain why Twitter users accept social TV. Providing a baseline to the uses and gratification theory assumptions, Technology Acceptance Model (TAM) explains that Social TV users are likelier to adopt and use the new journalistic approach when they perceive it gratifies their information-gathering and information-seeking needs. Consequently, it was proposed and validated that Social TV is preferred for Twitter users, who are already familiar with the platform and can participate in conversations using hashtags and mentions. The results remained consistent with the existing studies. For instance, the study by Athwal et al. [55] witnessed online media acceptance to gather Information about luxury brands. At the same time, Joo and Sang found mobile phone usage for news-gathering purposes among Korean youth [56].

Furthermore, this research also validated that Information seeking is important gratification linked to Social TV users' data journalism acceptance. These results are consistent with the study by Zhang et al. [57], as they considered user acceptance related to Information seeking among online consumers to search for relevant details, products and services. Gallego et al. [58] validated the relevant concept regarding online education purposes. Despite the third factor, "Convenience" remained insignificant, many studies [59, 60] consider it one of the primary factors linked to gratifications. As noted by Wang et al. [61], satisfaction and gratification are long-standing phenomena when technology acceptance is questioned. However, these factors are linked with the approaches they have and the effort they apply. Consequently, the more they feel convenience, the more they will likely accept.

Further, the effect of gratifications on Social TV users was found significantly affect their data journalism acceptance. These findings are in line with the propositions by [22] Erduran and Demirkiran as they argued, Data journalism gratifies users' needs by providing detailed and subtle insights into complex issues. Consequently, Social TV users accept it as they often seek reliable and informative content to share with their followers. Finally, Perceived Usefulness also remained indirectly affecting the acceptance behaviour, indicating the applicability of the Technology Acceptance Model on data journalism, adding more significance to current research. As Koehler & Parrell [62] argued, expected Usefulness is important in social TV users' acceptance of data journalism. As data journalism can be easily understood and engaged through visually compelling data representation, it positively affects the users' acceptance and perceptions [63].

6   Conclusion and Limitations

Current research study has shown that data journalism gratifies the information needs of social TV users, ultimately contributing to its acceptance in the United Arab Emirates. Today, data journalism has become a valuable source of information for social TV users as it provides Twitter users with detailed insights into complicated issues in an easily understandable and engaging format. Besides, perceived ease of use remains an important factor in social TV users' acceptance of data journalism, as it contributes to their willingness to engage with this content. Thus, as the social TV audience evolves from Twitter users to other platforms, it is usefully integrated with data journalism. Additionally, this study has some limitations. First, this study was conducted in the United Arab Emirates, which limits its generalizability in other regions. Second, the focus remained only on Twitter users, while social TV is also popular among other social network users. Finally, the third limitation is based on selecting limited types of gratifications, narrowing the scope of current research.

References

1. Tong, J.: Revisiting the importance of data journalism. In: Journalism, Economic Uncertainty and Political Irregularity in the Digital and Data Era, pp. 47–60. https://doi.org/10.1108/978-1-80043-558-220221004
2. Appelgren, E., Lindén, C.-G.: Data journalism as a service: digital native data journalism expertise and product development. Media Commun. **8**(2), 62–72 (2020). https://doi.org/10.17645/mac.v8i2.2757
3. Zhang, X., Chen, M.: Journalists' adoption and media's coverage of data-driven journalism: a case of Hong Kong. Journal. Pract. **16**(5), 901–919 (2022). https://doi.org/10.1080/17512786.2020.1824126

4. Camaj, L., Martin, J., Lanosga, G.: The impact of public transparency infrastructure on data journalism: a comparative analysis between information-rich and information-poor countries. Digit. Journal., 1–20 (2022). https://doi.org/10.1080/21670811.2022.2077786

5. Porlezza, C., Splendore, S.: From open journalism to closed data: data journalism in Italy. Digit. Journal. 7(9), 1230–1252 (2019). https://doi.org/10.1080/21670811.2019.1657778

6. Bebawi, S.: Data journalism and investigative reporting in the arab world: from emotive to evidence-based journalism. In: Mutsvairo, B., Bebawi, S., Borges-Rey, E. (eds.) Data Journalism in the Global South. PSJGS, pp. 193–204. Springer, Cham (2019). https://doi.org/10.1007/978-3-030-25177-2_11

7. European Journalism Centre: Why Is Data Journalism Important? DataJournalism.com (2023). https://datajournalism.com/read/handbook/one/introduction/why-is-data-journalism-important. Accessed 07 Apr 2023

8. Baloch, K., Andreson, K.: Reporting in conflict zones in Pakistan: risks and challenges for fixers. Social Science Open Access Repository (SSOAR) (2020). https://www.ssoar.info/ssoar/handle/document/66735. Accessed 05 Mar 2023

9. Farhi, F., Saidani, S., Jeljeli, R.: Determinants of dependence on social media for accessing news a study on a sample of journalists in the newsroom. Psychol. Educ. 58, 3531–3543 (2021)

10. Mazouz, A., Alnaji, L., Jeljeli, R., Al-Shdaifat, F.: Innovation indicators and growth in the United Arab Emirates. AAU J. Bus. Law 1, 1–12 (2017). https://doi.org/10.51958/AAUJBL 2017V1I1P2

11. Bravo, A.A., Tellería, A.S.: Data journalism: from social science techniques to data science skills. Hipertext.net (2020). https://doi.org/10.31009/hipertext.net.2020.i20.04

12. Widholm, A., Appelgren, E.: A softer kind of hard news? Data journalism and the digital renewal of public service news in Sweden. New Media Soc. 24(6), 1363–1381 (2022). https://doi.org/10.1177/1461444820975411

13. Auväärt, L.: Fighting COVID-19 with data: an analysis of data journalism projects submitted to sigma awards 2021. Cent. Eur. J. Commun. 15(3(32)), 379–395 (2023). https://doi.org/10.51480/1899-5101.15.3(32).3

14. Desai, A., Nouvellet, P., Bhatia, S., Cori, A., Lassmann, B.: Data journalism and the COVID-19 pandemic: opportunities and challenges. Lancet Digit Health 3(10), e619–e621 (2021). https://doi.org/10.1016/S2589-7500(21)00178-3

15. Lewis, N.P., Nashmi, E.A.: Data journalism in the Arab region: role conflict exposed. Digit. Journal. 7(9), 1200–1214 (2019). https://doi.org/10.1080/21670811.2019.1617041

16. Fahmy, N., Attia, M.A.: A field study of Arab data journalism practices in the digital era. Journalism Pract. 15(2), 170–191 (2021). https://doi.org/10.1080/17512786.2019.1709532

17. Snoussi, T.: ICT faculties' usage in the UAE private universities: a case study. Glob. Media J. 17(33), 1–8 (2019)

18. Kabha, R.: Comparison study between the UAE, the UK, and India in dealing with WhatsApp fake news. JCCC 10(9), 176–186 (2019). https://doi.org/10.31620/JCCC.12.19/18

19. Appelgren, E., Lindén, C.-G., van Dalen, A.: Data journalism research: studying a maturing field across journalistic cultures, media markets and political environments. Digit. Journal. 7(9), 1191–1199 (2019). https://doi.org/10.1080/21670811.2019.1685899

20. deLimaSantos, M.-F., Mesquita, L.: Data journalism beyond technological determinism. Journal. Stud. 22(11), 1416–1435 (2021). https://doi.org/10.1080/1461670X.2021.1944279

21. Arias-Robles, F., López, P.J.L.: Driving the closest information. local data journalism in the UK. Journal. Pract. 15(5), 638–650 (2021). https://doi.org/10.1080/17512786.2020.1749109

22. Erduran, Y., Demirkiran, T.: Are 'Data Journalism Trainings' effective in building participatory journalism? (An example from Turkey). Journalism Practice, 13 (2019)

23. Allern, S., Pollack, E.: Journalism as a public good: a Scandinavian perspective. Journalism 20(11), 1423–1439 (2019). https://doi.org/10.1177/1464884917730945

24. Geng, S., Law, K.M.Y., Niu, B.: Investigating self-directed learning and technology readiness in blending learning environment. Int. J. Educ. Technol. High. Educ. **16**(1), 1–22 (2019). https://doi.org/10.1186/s41239-019-0147-0

25. Mitchell, A., Holcomb, J., Page, D.: Investigative journalists and digital security. Pew Research Center (2015). https://www.pewresearch.org/journalism/wp-content/uploads/sites/8/2015/02/PJ_InvestigativeJournalists_020515.pdf

26. DataStories: Data Journalism. Data Stories (2022). https://www.datastories.pk/. Accessed 05 Mar 2023

27. Loosen, W., Reimer, J., De Silva-Schmidt, F.: Data-driven reporting: an on-going (r)evolution? An analysis of projects nominated for the data journalism awards 2013–2016. Journalism **21**(9), 1246–1263 (2020). https://doi.org/10.1177/1464884917735691

28. Showkat, D., Baumer, E.P.S.: Where do stories come from? Examining the exploration process in investigative data journalism | proceedings of the ACM on human-computer interaction. In: Proceedings of the ACM on Human-Computer Interaction (2021). https://doi.org/10.1145/3479534

29. Tandoc, E., Hess, K., Eldridge, S., Westlund, O.: Diversifying diversity in digital journalism studies: reflexive research, reviewing and publishing. Digit. Journal. **8**(3), 301–309 (2020). https://doi.org/10.1080/21670811.2020.1738949

30. Hofmann, T., et al.: Technology readiness and overcoming barriers to sustainably implement nanotechnology-enabled plant agriculture. Nature Food **1**(7), 416–425 (2020). https://doi.org/10.1038/s43016-020-0110-1

31. Mutsvairo, B.: Challenges facing development of data journalism in non-western societies. Digit. Journal. **7**(9), 1289–1294 (2019). https://doi.org/10.1080/21670811.2019.1691927

32. Figueiras, M.J., Ghorayeb, J., Coutinho, M.V.C., Marôco, J., Thomas, J.: Levels of trust in information sources as a predictor of protective health behaviors during COVID-19 pandemic: a UAE cross-sectional study. Frontiers in Psychology **12** (2021). https://www.frontiersin.org/articles/10.3389/fpsyg.2021.633550. Accessed 14 Jan 2023

33. Camilleri, M.A., Falzon, L.: Understanding motivations to use online streaming services: integrating the Technology Acceptance Model (TAM) and the Uses and Gratifications Theory (UGT). Spanish Journal of Marketing - ESIC **25**(2), 217–238 (2020). https://doi.org/10.1108/SJME-04-2020-0074

34. Choi, E.-K., Fowler, D., Goh, B., Yuan, J.: Social media marketing: applying the uses and gratifications theory in the hotel industry. J. Hosp. Mark. Manage. **25**(7), 771–796 (2016). https://doi.org/10.1080/19368623.2016.1100102

35. Dolan, R., Conduit, J., Fahy, J., Goodman, S.: Social media engagement behaviour: a uses and gratifications perspective. J. Strateg. Mark. **24**(3–4), 261–277 (2016). https://doi.org/10.1080/0965254X.2015.1095222

36. Jeljeli, R., Farhi, F., Hamdi, M.E., Saidani, S.: The impact of technology on audiovisual production in the social media space. Acad. J. Interdiscip. Stud. **11**(6), 48 (2022). https://doi.org/10.36941/ajis-2022-0148

37. Pasha, S.A., Youssef, E., Sharif, H.: Role of virtual reality in improving students' LMS experiences: structural equation modelling based study. In: 2021 International Conference of Modern Trends in Information and Communication Technology Industry (MTICTI), pp. 1–7 (2021). https://doi.org/10.1109/MTICTI53925.2021.9664769

38. de-Lima-Santos, M.-F., Mesquita, L.: Data journalism in favela: made by, for, and about forgotten and marginalized communities. Journalism Practice, 1–19 (2021). https://doi.org/10.1080/17512786.2021.1922301

39. Saboune, F.M.F.: Virtual reality in social media marketing will be the new model of advertising and monetization. In: IEEE Conference Publication I IEEE Xplore, presented at the International Conference on Social Networks Analysis, Management and Security (SNAMS) (2022). https://ieeexplore.ieee.org/abstract/document/10062551. Accessed 02 Apr 2023

40. Farhi, F., Jeljeli, R., Zahra, A., Saidani, S., Feguiri, L.: Factors behind virtual assistance usage among iPhone users: theory of reasoned action. Int. J. Interact. Mob. Technol. **17**(02), 42–61 (2023). https://doi.org/10.3991/ijim.v17i02.36021

41. Farhi, F., Chettah, M.: Social media as news sources for TV news channels -a study in the newsrooms. Talent Development & Excellence **12**(3) (2020)

42. Aljumah, A., Nuseir, M., Refae, G.: Examining the effect of social media interaction, E-WOM, and public relations: assessing the mediating role of brand awareness. Int. J. Data Netw. Sci. **7**(1), 467–476 (2023)

43. Islam, T., Mahmood, K., Sadiq, M., Usman, B., Yousaf, S.U.: Understanding knowledgeable workers' behavior toward COVID-19 information sharing through WhatsApp in Pakistan. Frontiers in Psychology 11 (2020). https://doi.org/10.3389/fpsyg.2020.572526. Accessed 18 Mar 2023

44. Aydin, G.: Examining social commerce intentions through the uses and gratifications theory. Int. J. E-Bus. Res. **15**(2), 44–70 (2019). https://doi.org/10.4018/IJEBR.2019040103

45. Luo, M.M., Chea, S., Chen, J.-S.: Web-based information service adoption: a comparison of the motivational model and the uses and gratifications theory. Decis. Support Syst. **51**(1), 21–30 (2011). https://doi.org/10.1016/j.dss.2010.11.015

46. Jeljeli, R., Farhi, F., Zahra, A.: Impacts of PR and AI on the reputation management: a case study of banking sector customers in UAE I springerprofessional.de. In: Digitalisation: Opportunities and Challenges for Business, Springer International Publishing (2022). https://www.springerprofessional.de/impacts-of-pr-and-ai-on-the-reputation-management-a-case-study-o/24076546. Accessed 18 Mar 2023

47. Jeljeli, R., Farhi, F., Hamdi, M.E.: The impact of technology on audiovisual production in the social media space. Acad. J. Interdisc. Stud. (2022). https://www.richtmann.org/journal/index.php/ajis/article/view/13099. Accessed 18 Mar 2023

48. Global Media Insights: United Arab Emirates (UAE) Social Media Statistics 2023 I GMI (2023). https://www.globalmediainsight.com/blog/uae-social-media-statistics/. Accessed 25 Apr 2023

49. Krejcie, R.V., Morgan, D.W.: Determining sample size for research activities. Educ. Psychol. Measur. **30**(3), 607–610 (1970). https://doi.org/10.1177/001316447003000308

50. Aljumah, A.I., Nuseir, M.,T., Refae, G.A.E.: The effect of sensory marketing factors on customer loyalty during Covid 19: exploring the mediating role of customer satisfaction. **6**(4), 1359–1368 (2022). https://doi.org/10.5267/j.ijdns.2022.5.015

51. Asthana, A.: Multicollinearity (2018)

52. Jeljeli, R., Farhi, F., Zahra, A.: Impacts of PR and AI on the reputation management: a case study of banking sector customers in UAE. In: Alareeni, B., Hamdan, A., Khamis, R., El Khoury, R. (eds.) Digitalisation: Opportunities and Challenges for Business: Volume 1, pp. 265–277. Springer International Publishing, Cham (2023). https://doi.org/10.1007/978-3-031-26953-0_26

53. Demler, O.V., Paynter, N.P., Cook, N.R.: Tests of calibration and goodness-of-fit in the survival setting. Statist. Med. **34**(10), 1659–1680 (2015). https://doi.org/10.1002/sim.6428

54. Novak, A.R., Impellizzeri, F.M., Garvey, C., Fransen, J.: Implementation of path analysis and piecewise structural equation modelling to improve the interpretation of key performance indicators in team sports: an example in professional rugby union. **39**(22), 2509–2516 (2021). https://doi.org/10.1080/02640414.2021.1943169

55. Athwal, N., Istanbulluoglu, D., McCormack, S.E.: The allure of luxury brands' social media activities: a uses and gratifications perspective. Inf. Technol. People **32**(3), 603–626 (2018). https://doi.org/10.1108/ITP-01-2018-0017

56. Joo, J., Sang, Y.: Exploring Koreans' smartphone usage: an integrated model of the technology acceptance model and uses and gratifications theory. Comput. Hum. Behav. **29**(6), 2512–2518 (2013). https://doi.org/10.1016/j.chb.2013.06.002

57. Zhang, J., Shabbir, R., Mujeeb-ur-Rehman, A.: Role of social media in pre-purchase consumer information search: a uses and gratifications perspective. Mediterr. J. Soc. Sci. **6**(1), 11 (2015)

58. Gallego, M.D., Bueno, S., Noyes, J.: Second life adoption in education: a motivational model based on uses and gratifications theory. Comput. Educ. **100**, 81–93 (2016). https://doi.org/10.1016/j.compedu.2016.05.001

59. Gamage, T.C., Tajeddini, K., Tajeddini, O.: Why Chinese travelers use WeChat to make hotel choice decisions: a uses and gratifications theory perspective. J. Glob. Scholars Market. Sci. **32**(2), 285–312 (2022). https://doi.org/10.1080/21639159.2021.1961599

60. Ray, A., Dhir, A., Bala, P.K., Kaur, P.: Why do people use food delivery apps (FDA)? A uses and gratification theory perspective. J. Retail. Consum. Serv. **51**, 221–230 (2019). https://doi.org/10.1016/j.jretconser.2019.05.025

61. Wang, C., Teo, T.S.H., Dwivedi, Y., Janssen, M.: Mobile services use and citizen satisfaction in government: integrating social benefits and uses and gratifications theory. Inf. Technol. People **34**(4), 1313–1337 (2021). https://doi.org/10.1108/ITP-02-2020-0097

62. Koehler, S.N., Parrell, B.R.: The impact of social media on mental health: a mixed-methods research of service providers' awareness. California State University, San Bernardino, United States (2020). https://scholarworks.lib.csusb.edu/cgi/viewcontent.cgi?article=2131&context=etd

63. Burgess, J.: The SAGE handbook of social media. The SAGE Handbook of Social Media, pp. 1–662 (2017)

64. Lowrey, W., Broussard, R., Sherrill, L.A.: Data journalism and black-boxed data sets. Newsp. Res. J. **40**(1), 69–82 (2019). https://doi.org/10.1177/0739532918814451

Unveiling the Performance Insights: Benchmarking Anomaly-Based Intrusion Detection Systems Using Decision Tree Family Algorithms on the CICIDS2017 Dataset

Mohamed Azalmad(✉), Rachid El Ayachi(iD), and Mohamed Biniz

Laboratory TIAD, Department of Computer Science, Sultan Moulay Slimane University, Beni Mellal, Morocco
azalmad.mohamed@gmail.com, rachid.elayachi@usms.ma

Abstract. The continuous growth of computer networks and the internet has brought attention to the increasing potential damage caused by attacks. Intrusion Detection Systems (IDSs) have emerged as crucial defense tools against the rising frequency and sophistication of network attacks. However, effectively detecting new attacks using machine-learning approaches in intrusion detection systems presents challenges.

This study focuses on the CICIDS2017 dataset, which is one of the most recent and updated IDS datasets publicly available. The CICIDS2017 dataset contains both benign and seven common attack network flows, meeting real-world criteria and providing true network traffic data. Furthermore, The CICIDS2017 dataset presents challenges when it comes to measuring the performance of a comprehensive set of machine learning algorithms in order to identify the optimal pattern set for detecting specific attack categories.

This paper contributes to the field of intrusion detection systems by benchmarking the decision tree family. As a result of our study XGBoost achieves the highest accuracy of 99%, followed by Random Forest with 98%, Gradient Boosting Trees with 88%, and Decision Tree with 89%.

Overall, this research provides valuable insights into the performance of decision tree family and feature selection methods, paving the way for the advancement of more reliable and efficient intrusion detection systems.

Keywords: CICIDS-2017 · Decision tree family algorithms · IDs · Decision tree · Random Forest · Gradient Boosting · eXtreme Gradient Boosting

1 Introduction

In recent years, we have heard a lot about the term "cybersecurity," which encompasses several details to discover. The world of technology is constantly and rapidly evolving to meet human needs, both as individuals and as organizations, as well as to address security threats. Nowadays, everyone uses developed computer systems, allowing us to process and transmit data very quickly. In recent years, the amount of exchanged and

R. El Ayachi et al. (Eds.): CBI 2023, LNBIP 484, pp. 202–219, 2023.
https://doi.org/10.1007/978-3-031-37872-0_15

stored data has also increased. All this sensitive data must be secured and protected, and this is where the role of intrusion detection systems (IDS) comes into play. IDS includes various methods and techniques used to define and protect data and systems on the network against security threats.

An IDS is actually a vital function within the cybersecurity framework. Its main objective is to analyze the network and detect unusual events in real-time. An intrusion detection system can vary from one system to another based on four main components: (1) the position of the IDS on the network, (2) the detection method used by the IDS, (3) the behavior after detection, and (4) the architecture of the IDS. Each component has more than one approach behind it, although each IDS can have only one approach for each component [1].

In the case of our study, the most important component is "the approach used by the IDS to detect unusual events." IDS can have multiple approaches to deal with intrusions, but the approach we will focus on in this article is the machine learning-based approach. The volume of traffic circulating on networks is significant, giving network-connected systems the ability to collect and store a massive amount of data in large databases. This availability and wealth of data are leveraged by machine learning-based intrusion detection systems to find and detect abnormal events and behaviors that compromise the overall system's security. The enormous amount of data in recent years has enabled intrusion detection systems based on machine learning techniques to be more powerful and intelligent than traditional approaches and techniques [2].

The rest of the paper is organized as follows: Sect. 2 gives some works related to the subject treated, Sect. 3 presents an Overview of intrusion detection systems, Sect. 4 shows the methodology adopted to benchmark machine learning models, Sect. 5 is reserved for experimental results and the conclusion is given in Sect. 6.

2 Related Works

Recently, there has been significant interest in applying machine learning techniques to intrusion detection systems. Many research works implement different approaches to automate the discovery process of attacks and regular events in massive datasets. This section will discuss the different perspectives proposed to deal with intrusion detection based on machine-learning approaches and summarize the recent research works.

Doaa N. Mhawi et al. [3] propose a novel approach to build four classifiers on the CICIDS2017 dataset with high accuracy, low false alarm rates, low false negative rates, and better reliability. The proposed approach consists of three main phases. The first phase is preprocessing, which removes duplicate data, transforms categorical attributes into numerical using Label-Encoder and One-Hot Encoding techniques, and normalises all the features in the dataset based on the min-max function. The second phase consist of reducing the dimensionality of the dataset and selecting the best 30 features to increase the performance based on the Correlation Feature Selection-Forest Panelized Attribute (CFS-FPA) method. The last third phase aims to build a Hybrid ensemble learning algorithm with four different classifiers (SVM, RF, NB, and KNN) based on the Adaboosting Algorithm as the first step and then the bagging Algorithm as the second step. The proposed hybrid ensemble learning method with the CFS-FPA approach for feature selection attains an accuracy of 99.73%, an F-measure of 99.71%, a precision of 99.82%, a DR of 99.8%, and a FAR of 0.004.

Ziadoon Kamil Maseer et al. [4] benchmarked seven estimators and three unsupervised ML algorithms to investigate the ability of these algorithms to detect attacks and normal behaviors on the CICIDS2017 dataset. The collection of estimators includes the artificial neural network (ANN), decision tree (DT), k-nearest neighbor (k-NN), naive Bayes (NB), random forest (RF), support vector machine (SVM), and convolutional neural network (CNN) algorithms. In contrast, unsupervised ML algorithms include expectation-maximization (EM), k-means, and self-organizing maps (SOM) algorithms. The proposed benchmarking methodology does not implement any technique to handle the high-class imbalance issue. Instead, it implements the feature selection by selecting 38 features with values in the range of $[-3, 3]$ after standardizing all the features and manually fine-tuning hyper-parameters for all the models. The benchmarking results show that the supervised models NB, KNN, and DT obtain the best results compared with other models, with the max accuracy value of 0.9886% obtained by NB while the min accuracy value of 0.7521% obtained by the SVM model. While for the unsupervised models, the accuracy value does not exceed 0.6006%.

Arif Yulianto et al. [5] propose an approach to improve the AdaBoost performance on the CICIDS2017 dataset. The proposed approach consists of implementing two techniques for feature selection: Ensemble Feature Selection (EFS) and Principal Component Analysis (PCA). Alongside the SMOTE technique to handle the high-class imbalance problem. The research work benchmarked the performance metric of the following four combinations of the proposed techniques: AdaBoost + EFS, AdaBoost + EFS + SMOTE, AdaBoost + PCA, and AdaBoost + PCA + SMOTE. The AdaBoost-based Intrusion Detection System (IDS) with the EFS and SMOTE techniques reaches the highest performance with an accuracy of 81.83%, recall of 100%, and F1 Score of 90.01% compared to the other combinations. While in the other hand, the Adaboost + EFS without using the SMOT technique reaches the highest precision of 85.15%.

Kurniabudi [6] provide a study to prove the impact of feature selection in improving attack detection and time execution in the CICIDS 2017 dataset based on the information gain technique. The author uses the Information Gain technique to assign a weight to each future and then build 6 groups of features based on the score of the feature's weight. Then implement Naive Bayes (NB), Bayes Net (BN), Random Tree (RT), J48, and Random Forest (RF) classifiers on every group of futures. These futures groups include the 4-features group, 15-features group, 22-features group, 35-features group, 52-features group, 57-features group, and all-features group. The Table below summarizes the best experiment result obtained by a group of features for each classifier alongside the time execution (Table 1).

Table 1. Performance Comparison of Different Classifiers with Varied Feature Sets

Classifier	Group of features	Accuracy	FPR	Time execution(s)
J48	All-features	0.9988	0.002	2289
RF	22-features	0.9986	0.003	2733
RT	15-features	0.9979	0.005	38
BN	4-features	0.9526	0.010	35
NB	35-features	0.7084	0.013	50

3 Overview of Intrusion Detection Systems

In recent years, the significance of intrusion detection systems (IDS) has grown substantially, driven by the escalating frequency and severity of network attacks. IDS has become an indispensable component of the security infrastructure for numerous organizations. Its primary role involves continuous monitoring and identification of intrusions within network events. As defined in [2], an intrusion encompasses any malicious activity aimed at compromising the integrity, confidentiality, or availability of system resources [11]. IDS can be categorized into various groups based on different characteristics [10]. Figure 1 illustrates the principal functional components used for classifying IDS [16,17].

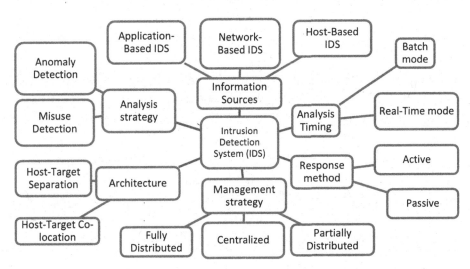

Fig. 1. Taxonomy of an intrusion detection system

4 Methodology

We propose an experimental process to benchmark decision tree family algorithms learning and achieve maximum accuracy. In this section, we will present and describe the dataset we worked with. Additionally, we will discuss the techniques employed to prepare the dataset for training and evaluation, including preprocessing. These techniques played a vital role in maximizing the performance metrics of the proposed models. Furthermore, we will delve into the selected models and the techniques used to fine-tune their parameters. Finally, we will outline the evaluation metrics employed to assess the performance of these models. The proposed system consists of four main stages, as illustrated in the figure below, which will be detailed in the subsequent sections (Fig. 2):

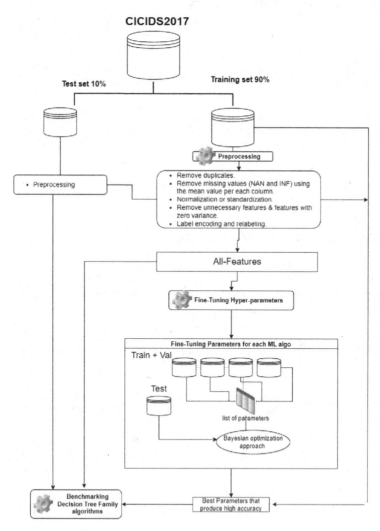

Fig. 2. Proposed system of intrusion detection

We split the dataset into training and test datasets, with the training dataset accounting for approximately 90% of the data and the test dataset for the remaining 10%. To prevent information leaks from the test dataset to the training dataset, we perform preprocessing on the training dataset. We then apply these measures to the test dataset without involving the test dataset values in the preprocessing phase.

The preprocessing stage involves cleaning the dataset by removing duplicates, converting categorical features into numerical values, relabeling the target class label with numeric values, filling in missing values, and normalizing the dataset based on its columns.

In the subsequent stage, we proceed to fine-tune the hyperparameters of the decision tree family classifiers: decision trees (DT), random forest (RF), Gradient Boosting (GB), and eXtreme Gradient Boosting (XGBoost). The primary objective of this stage is to optimize the hyperparameters for each machine learning algorithm using the Bayesian approach, which enables us to identify the optimal hyperparameters more efficiently compared to traditional grid or pure randomized search approaches. From the range of available parameter options, we select the most critical set of parameters. By training the algorithms on the parameter space, we determine the best combination of hyperparameters that yield the highest accuracy. Finally, we train each machine learning algorithm using the optimized parameters, thereby creating models that demonstrate exceptional performance for future predictions.

4.1 Dataset

Datasets play a crucial role in the training and testing of models. A high-quality dataset is essential for researchers to develop robust models and ensure effective validation and testing. In the field of intrusion detection systems, there are various publicly available datasets. However, many of these datasets are outdated and lack the representation of the latest attacks, while others fail to reflect the real state of networks. To address these concerns, we established two key criteria for selecting a dataset: (1) the presence of up-to-date common attacks and (2) the availability of true network traffic data. The CICIDS2017 dataset perfectly aligns with our criteria, and we have chosen it for benchmarking the selected machine learning models.

The CICIDS2017 dataset, generated in 2017 by the Canadian Institute for Cybersecurity, consists of realistic network traffic data captured using the CICFlowMeter network traffic analysis tool. It encompasses both benign behaviors and the newest attacks targeting HTTP, HTTPS, FTP, SSH, and email protocols, making it representative of real-world data [9].

Furthermore, the CICIDS2017 dataset meets the criteria proposed by the evaluation framework for a valid IDS dataset. These criteria ensure that a valid IDS dataset possesses characteristics such as attack diversity, anonymity, available protocols, complete capture, complete interaction, complete network configuration, complete traffic, feature set, heterogeneity, labeling, and metadata.

The data in the CICIDS2017 dataset was captured over five workdays, spanning from 9 a.m. on Monday, July 3, 2017, to 5 p.m. on Friday, July 7, 2017. It includes both benign traffic and common attack types such as DoS, DDoS, brute force, XSS, SQL injection, infiltration, port scan, and botnet activities [7]. The dataset is divided into eight CSV files, with each file differing in traffic volume and labels. The table below provides a summary of the traffic volume, class labels, and distribution of instances in each CSV file [8] (Table 2).

To build the desired machine learning models, we extract the dataset from each file and combine all these sub-datasets into one extensive dataset. The obtained dataset becomes enormous, consisting of 2,830,743 rows, 79 features, and comprising 15 class labels, a single class label representing the normal behavior named BENING, and the remaining 14 labels representing the attacks, including, Bot, DDoS, DoS GoldenEye, DoS Hulk, DoS Slowhttptest, DoS slow loris, FTP-Patator, Heartbleed, Infiltration, PortScan, SSH-Patator, Web Attack/Brute Force, Web Attack/SQL Injection, Web Attack/XSS. The table below shows the number of instances per class label [6] (Table 3).

The first stage in the benchmarking process consists of splitting the dataset into training datasets with 80% of instances, and the testing dataset, with the rest. Working with datasets directly without any pre-treatment is not suitable for model performance. The training CICIDS2017 dataset, like many other datasets, faces three major issues that must be handled carefully before starting the training and testing phases. The first issue is redundant data, incomplete, missing values, and features with different scales. The second issue is the high dimensionality of the dataset. CICIDS2017 consists of 79 variables; not all variables are helpful and vital to building the models. There are redundant and irrelevant features that lead to reduced generalization capability of the model and the overall accuracy. The third issue is the class imbalance. as shown in the Table above, CICIDS2017 has a significant imbalance between the normal class and observation in other classes. Also, there is a significant imbalance between the attack classes. Some types of attacks are represented with a considerable number of instances, while others have small instances.

Before we get directly into the training and testing phases, we need to implement preprocessing, feature selection, and class imbalance techniques to optimize the dataset and make it 100% valid to reach the maximum accuracy performance. All these techniques will be discussed in detail in the following sections.

4.2 Preprocessing

Preprocessing is a vital task to enhance data quality. Training machine learning models with such bad quality data will generally produce a bad model at generalization. CICIDS2017, like any other real-world dataset, contains null values, duplicates, outliers, and features with different scales [8]. We implement a set of steps to prepare the raw data to make it suitable for benchmarking the machine learning algorithms.

The first step consists of deleting all the duplicate data, with the condition that the duplicates should have the same values on all the columns. Two rows with the same values on all the columns are considered duplicates; otherwise, they are considered unique. The CICIDS2017 dataset consists of 3,119,345 records, with a total of 288,804 duplicates, which represent 9.25% of the total number of instances. We are removing duplicates only from the training dataset and not considering the testing dataset in the preprocessing stage.

In the imputer step, we address the issue of missing values by replacing them with a statistical value for each column. Several approaches can be used to select the statistical

Table 2. Details of CSV files.

Days	File name	Traffic volume	Labels	Traffic volume/Labels	
Monday, July 3, 2017	Monday-WorkingHours.pcap ISCX.csv	529,918	BENIGN	BENIG	529,918
Tuesday, July 4, 2017	Tuesday-WorkingHours.pcap_ISCX.csv	445,909	BENIGN, FTP-Patator, SSH-Patator	BENIGN	432,074
				FTP-Patator	7938
				SSH-Patator	5897
Wednesday, July 5, 2017	Wednesday-workinHours.pcap_ISCX.csv	692,703	BENIGN, DoS GoldenEye', DoS Hulk, DoS Slowhttptest,DoS slowloris, Heartbleed	BENIGN	440,031
				DoS GoldenEye	10,293
				DoS Hulk	231,073
				DoS Slowhttptest	5499
				DoS slowloris	5796
				Heartbleed	11
Thursday, July 6, 2017	Thursday-WorkingHours-Morning-WebAttacks.pcap_ISCX.csv	458,968	BENIGN,Web Attack Brute Force, Web Attack Sql Injection, Web Attack XSS	BENIGN	168,186
				Web Attack Brute Force	1507
				Web Attack Sql Injection	21
				Web Attack XSS	652
	Thursday-WorkingHours-Afternoon-Infilteration.pcap_ISCX.csv	288,602	BENIGN, Infiltration	BENIGN	288,566
				Infiltration	36
Friday, July 7, 2017	Friday-WorkingHours-Morning.pcap_ISCX.csv	191,033	BENIGN, Bot	BENIGN	189,067
				Bot	
	Friday-WorkingHours-Afternoon-PortScan.pcap_ISCX.csv	286,467	BENIGN, PortScan	BENIGN	127,537
				PortScan	158,930
	Friday-WorkingHours-Afternoon-DDos.pcap_ISCX.csv	225,745	BENIGN DDos	BENIGN	97,718
				DDos	128,027

Table 3. Number of instances per class label.

Class label	Number of instances	Percentage %
BENING	2,273,097	80.0%
DoS Hulk	231,073	8.162%
PortScan	158,930	5.614%
DoS GoldenEye	128,027	4.522%
DoS Hulk	10,293	0.363%
FTP-Patator	7,938	0.280%
SSH-Patator	5,897	0.208%
DoS slowloris	5,796	0.204%
DoS Slowhttptest	5,499	0.194%
Bot	1,966	0.069%
Web Attack/Brute Force	1,507	0.053%
Web Attack/XSS	652	0.023%
Infiltration	36	0.0012%
Web Attack/SQL Injection	21	0.0000741%
Heartbleed	11	0.0000388%

value, such as mean, median, mode, most frequent value, nearest neighbor value (KNN), or a constant. The CICIDS2017 dataset contains a significant number of missing values, with 24,536,904 marked as NaN and Inf. To ensure accurate benchmarking of our collection of machine learning algorithms, we chose to replace the missing values with the mean value. We calculated the mean value for each column based solely on the training dataset and then replaced all missing values in both the training and testing datasets with their respective mean values.

Feature normalization is a crucial step in enhancing and facilitating the learning phase of classifiers that rely on gradient descent and distance calculations, such as gradient boosting and XGBoost. The CICIDS2017 dataset consists of features with different scales, ranging from extensive ranges to small ranges, as well as features with significant values and others with small values.

To address this, we rescale the features in our dataset using the standardization formula. This formula transforms the features to a similar range by centering each feature value, subtracting the mean, and dividing by the standard deviation. This process shifts the distribution to have a mean of zero and a standard deviation of one.

Lastly, we handle categorical features by converting them into numeric values. Most of our selected machine learning algorithms, except for decision trees and random forests, are unable to handle categorical values directly. The CICIDS2017 dataset does not contain many categorical features, except for the target attribute, which includes 15 unique categorical values. To address this, we implement label encoding, assigning an integer value to each unique category.

4.3 Decision Tree Family Algorithms

In the machine learning era, several approaches have been proposed to detect attacks in large databases, each with its strengths and weaknesses. In our research, we work with the decision tree family algorithms. Our main goal is to train and test these algorithms on the CICIDS2017 dataset, then make an analysis study of performance for each machine learning model and compare the obtained results.

Decision Trees (DT)

The idea behind decision trees for classification is to split the data into subsets where each subset belongs to approximately only one class. The goal is to divide the initial given data into subsets that are as pure as possible, with the condition that each subset contains as many samples as possible from a single class. The decision tree is a hierarchical structure with nodes and directed edges. There are three types of nodes: a root node at the top, leaf nodes at the bottom, and internal nodes. The root and internal nodes have test conditions, while each leaf node has a class label. The depth of a decision tree is the number of edges in the longest path from the root node to the leaf node, while the size of a decision tree is the total number of nodes in the tree [12].

The algorithm for constructing a decision tree model is called an induction algorithm. It repeatedly partitions data into successively purer subsets until some stopping criteria have been satisfied. The partition process determines which feature should be selected as a condition for splitting the samples. This split condition is based on an impurity measure to determine the goodness of a split. On the other hand, the stopping criteria process tests whether all the samples belong to the same class label or the same feature values or if the number of samples has fallen below some minimum threshold.

Random Forest (RF)

Many research papers on Anomaly-Based Intrusion Detection Systems have investigated the utilization of Random Forest algorithms, include [6 12, 13]. Random forest is an ensemble learning model. It combines multiple individual decision trees to produce an aggregate model that is more powerful than any single decision tree model. Each tree in the random forest model is built from a different random sample of the data called the bootstrap sample. The building process for each tree is almost the same as for a standard decision tree but with one crucial difference. The test split is calculated among a subset of features instead of all available features. The random forest model is quite sensitive to a parameter called max features. This parameter controls the number of features that must be randomly selected in each subset. If the max features parameter is so close to one, then the trees in the forest will likely be very different from each other and more complex. Otherwise, if the Max features parameter is close to the total number of features, then the trees in the forest will likely be similar and quite simpler. The trained random forest model predicts the target value for new instances by first predicting every tree in the random forest. For regression, the overall prediction is typically the mean of the individual tree predictions. While for classification, the overall prediction is based on a weighted vote. Each tree gives a probability for each possible target class label, averages the obtained probabilities across all the trees, then selects the class label with the highest probability. Many research papers based on anomaly detection used Random Forest includes.

Gradient Boosting Trees (GB)

Gradient-boosted decision trees are another tree-based ensemble method. Like the random forest, gradient-boosted trees used an ensemble of multiple trees to create more powerful prediction models for classification and regression, rather than the random forest that combines several random trees in parallel. The gradient-boosted ensemble builds a series of weak decision trees stacked next to each other, where each tree model reduces the previous tree's produced errors. The number of trees and the error rate have an inverse relationship; the increase in the number of trees leads to a decrease in the error rate and vice versa. There are two essential parameters to build any gradient-boosted tree, the number of estimators that controls the model complexity and the learning rate, which emphasizes each estimator to reduce the error rate of its predecessor.

eXtreme Gradient Boosting (XGBoost)

XGBoost, short for eXtreme Gradient Boosting, is an advanced machine learning algorithm that has gained popularity for its exceptional performance in various domains, including classification and regression tasks. It is an optimized implementation of the gradient boosting framework that combines the strengths of boosting algorithms and gradient descent optimization techniques.

At its core, XGBoost employs a boosting ensemble technique, where a sequence of weak prediction models, typically decision trees, are combined to form a powerful predictive model. Unlike traditional gradient boosting methods, XGBoost introduces several innovations to enhance model performance and efficiency.

One key innovation is the adoption of a novel regularization technique called "regularized learning objective." It incorporates penalties and constraints during the training process to control model complexity and prevent overfitting, resulting in improved generalization and robustness.

Another important feature of XGBoost is its ability to handle missing values in the dataset. By employing a learning algorithm that automatically learns how to best handle missing values, XGBoost reduces the need for manual preprocessing steps, making it a convenient choice for real-world datasets with missing data.

4.4 Hyper-parameters Optimization

In addition to the dataset pretreatment techniques, hyper-parameter optimization is another critical factor that impacts the estimators' performance. It aims to find the combination of hyper-parameters of a given estimator that generate the highest performance values. Running every combination from the range values for hyper-parameters takes a long time and slows down the optimization process, especially if the search space dimension is enormous [14]. For this reason, we work with an intelligent randomized search based on the Bayesian approach to identify the optimal hyper-parameters faster than the grid or pure randomized search approach.

The Bayesian optimization approach aims to find the hyper-parameters that minimize the error function of a model estimator without prior knowledge about its distribution

[13], as shown in the Algorithm below. The most important inputs to this Algorithm consist of a surrogate model F, an acquisition function A, an estimator objective function E, and a maximum number of iterations. The Algorithm recursively adds a pair value of (hyper-parameters, score) to an observation history H using the acquisition function and the estimator model [15]. Then recalculating the surrogate model based on the new values of the history H. The details of this Algorithm are explained below:

The algorithm also provides interpretable feature importance scores, allowing to gain insights into the relative importance of different features in the model's decision-making process. This information can be valuable for feature selection and understanding the underlying patterns driving the predictions.

INPUT: Surrogate model F, an acquisition function A, an estimator E, a maximum number of iterations N, a list of hyper-parameters X, and data D.

OUTPUT: list of pair values (hyper-parameters, score)

$H = \emptyset$

For i = 1 to N

Select a new hyper-parameters configuration by optimizing the acquisition function F:

$$x_i = argmax_{x \in X} A(x, F_{i-1})$$

Evaluate estimator E with newly obtained hyper-parameters configuration to obtain a new score:

$$s_i = E(x_i, D)$$

$H = H \cup (x_i, s_i)$

Train the surrogate model A_i using the latest history of samples H

return H

Two main steps are repeated until the max number of iterations is reached. First, the acquisition function generates a hyper-parameter x_i, which maximizes the acquisition function. Second, the Gaussian process or surrogate model is trained using the updated *history of (hyper-parameter, score) pair values*. In the end, the algorithm returns the history pair values containing the hyper-parameters that achieve the best score.

4.5 Evaluation Metrics for Benchmarking Machine Learning Models

Evaluation metrics are vital components to benchmark the machine learning models and explain the performance of each model. There are different model evaluation metrics, but we will discuss four metrics in this section: Accuracy, Precision, Recall, and F1-score.

The accuracy score is simply the proportion of the correct predictions. It is used to evaluate the model performance. However, this metric can be misleading, especially when the dataset has imbalanced outcomes; in this case, it will always predict with the majority class, which means consistently achieving a high accuracy score.

$$Accuracy = \frac{TP + TN}{Total} \tag{1}$$

Benchmarking machine learning models on the CICIDS2017 dataset using only the accuracy without considering other measures such as precision, recall, and F1 score is insufficient. Models with high accuracy and low precision or recall rate value could not be more helpful.

Precision is how good the models are at predicting the attacks. A low percentage rate means a high positive alarm, which also means a high chance that models predict attack events as normal. On the other hand, a high percentage rate means the models correctly predict the attack events.

$$Precision = \frac{TP}{TP + FP} \tag{2}$$

The recall is another critical metric for benchmarking models' performance on the CICIDS2017 dataset. It asses the models' ability to predict actual attacks. A shallow Recall value means that the models cannot predict the attacks and consider most of them ordinary, which could not be good, especially in the case of intrusion detection systems.

$$Recall = \frac{TP}{TP + FN} \tag{3}$$

The F1 score attempts to combine precision and recall into one metric.

$$F1\ Score = 2 * \frac{Precision * Recall}{Precision + Recall} \tag{4}$$

5 Experimental Setup, Results and Discussion

The evaluation of performance is a crucial aspect of any intrusion detection approach. Proper evaluation is essential to demonstrate the accuracy and viability of a detection mechanism in a real-time environment. Evaluating the accuracy and quality of a method or approach provides insight into its performance over time. As new vulnerabilities emerge over time, current evaluations may become outdated. Evaluating an Intrusion Detection System (IDS) involves activities such as collecting attack traces, establishing an appropriate IDS evaluation environment, and implementing robust evaluation methodologies. In this section, we present the performance evaluation metrics used to assess the decision tree family models on the CICIDS2017 dataset. These key metrics include accuracy, precision, recall, and F1 score.

5.1 Experimental Setup

The objective of this research paper is to evaluate the performance of decision tree family algorithms, aiming to enhance their reliability, achieve high accuracy, and high F1 score. The benchmarking process involves utilizing the CICIDS2017 dataset and employing binary forms of the confusion matrix to assess the algorithms' effectiveness. The experiments were carried out on Google Colab Pro, utilizing the following specifications: an Intel Xeon CPU @2.20 GHz, 13 GB RAM, a Tesla K80 accelerator, and 12 GB GDDR5 VRAM.

5.2 Results and Discussion

In this section, we present the results of our benchmarking study on the decision tree family algorithms, including Decision Tree, Random Forest, Gradient Boost, and XGBoost, using the CICIDS2017 dataset for intrusion detection. The evaluation aimed to assess the performance of these models in terms of accuracy, precision, Recall, and F1 score, along with the corresponding confusion matrix for each model.

Firstly, we construct a confusion matrix for each model after training and testing the decision tree family models on the CICIDS2017 dataset. The confusion matrix for each model is presented below in the corresponding tables (Tables 4, 5, 6 and 7):

Table 4. The results obtained from the evaluation of the Decision Tree (DT) model on the CICIDS2017 dataset are presented in the form of a confusion matrix.

		Predicted	
		Normal	Attack
Actual	Normal	226197	1323
	Attack	61	48023

Table 5. The results obtained from the evaluation of the Random Forest (RF) model on the CICIDS2017 dataset are presented in the form of a confusion matrix.

		Predicted	
		Normal	Attack
Actual	Normal	227293	227
	Attack	7246	48309

Table 6. The results obtained from the evaluation of the Gradient Boosting (GB) model on the CICIDS2017 dataset are presented in the form of a confusion matrix.

		Predicted	
		Normal	Attack
Actual	Normal	227473	180
	Attack	4520	51034

Firstly, we assessed the accuracy of each model. The Decision Tree model demonstrated a high accuracy of 99.49%, surpassing both the Gradient Boosting and Random Forest algorithms. While Random Forest utilizes ensemble learning to enhance overall accuracy, it achieved an accuracy of 97.36% compared to the Decision Tree. Gradient Boosting, with its iterative optimization approach, fell short of the Decision Tree

Table 7. The results obtained from the evaluation of the eXtreme Gradient boosting (XGBoost) model on the CICIDS2017 dataset are presented in the form of a confusion matrix.

		Predicted	
		Normal	Attack
Actual	Normal	226849	146
	Attack	61	84879

Table 8. Comparison of Accuracy Rates for Decision Tree, Random Forest, Gradient Boosting, and XGBoost Models in Intrusion Detection.

	Decision tree (DT)	Random Forest (RF)	Gradient Boosting (GB)	eXtreme Gradient Boosting (XGBoost)
Accuracy	99.49%	97.36%	98.34%	99.55%

model, achieving an accuracy of 98.34%. Remarkably, XGBoost displayed exceptional accuracy of 99.55%, highlighting its superior predictive performance (Table 8).

Then, we proceeded to calculate the precision of the models. Precision evaluates the proportion of true positives among the instances predicted as positive. In our case, we measured the accuracy of predicting attack events among instances predicted as normal. A low value of the precision metric indicates a high rate of false positive alarms. The Decision Tree model achieved a precision of 99.87%, while Random Forest exhibited a precision of 86.95%. Gradient Boosting did not outperform the Decision Tree model, achieving a precision of 91.86%, and XGBoost demonstrated the highest precision at 99.92%. These results underscore the models' effectiveness in accurately identifying and detecting attacks, thereby minimizing false negatives and enhancing the overall reliability of intrusion detection (Table 9).

Table 9. Comparison of Precision Rates for Decision Tree, Random Forest, Gradient Boosting, and XGBoost Models in Intrusion Detection.

	Decision tree (DT)	Random Forest (RF)	Gradient Boosting (GB)	eXtreme Gradient Boosting (XGBoost)
Precision	99.87%	86.95%	91.86%	99.92%

Next, we assessed the recall metric to evaluate the models' ability to identify true positives among the instances predicted as positive. Recall, also known as sensitivity or true positive rate, is a crucial measure in intrusion detection as it represents the models' effectiveness in correctly detecting attacks. We calculated the proportion of actual attacks correctly predicted among the instances predicted as attacks. The Decision Tree model achieved a recall of 97.31%, indicating that it correctly identified the actual attacks. Random Forest showed improved performance with a recall of 99.53%, demonstrating its ability to detect a higher proportion of attacks accurately. Gradient Boosting surpassed both models with a recall of 99.64%, exhibiting a high degree of sensitivity in detecting attacks. XGBoost achieved the highest recall at 99.82%, indicating its exceptional ability to identify and classify attacks accurately. These results underscore the models' effectiveness in detecting attacks and highlight their potential in enhancing the security of intrusion detection systems (Table 10).

Table 10. Comparison of Recall Rates for Decision Tree, Random Forest, Gradient Boosting, and XGBoost Models in Intrusion Detection.

	Decision tree (DT)	Random Forest (RF)	Gradient Boosting (GB)	eXtreme Gradient Boosting (XGBoost)
Recall	97.31%	99.53%	99.64%	99.82%

We also examined the F1 score, which considers both precision and recall. The Decision Tree model achieved an F1 score of 98.57%, while Random Forest exhibited an F1 score of 92.81%. Gradient Boosting surpassed the Random Forest model with an F1 score of 95.59%, showcasing its superior balance between precision and recall. XGBoost achieved the highest F1 score of 99.86%, further emphasizing its exceptional performance in detecting intrusions effectively.

Overall, the benchmarking results highlight the distinct characteristics and performance outcomes of each model in the decision tree family. The Decision Tree model stands out for its simplicity and relatively high accuracy compared to the ensemble-based models. However, Random Forest, Gradient Boosting, and XGBoost all exhibited notable accuracy, precision, and F1 scores, with XGBoost emerging as the top performer in our evaluation.

6 Conclusion

This paper presents a comprehensive benchmarking study of decision tree family models, including Decision Tree, Random Forest, Gradient Boost, and XGBoost, using the CICIDS2017 dataset for intrusion detection. The objective of the study was to evaluate and compare the performance of these models in terms of accuracy, precision, and F score. Through extensive experimentation and analysis, distinct characteristics and performance outcomes were observed for each model.

The Decision Tree model demonstrated simplicity and achieved high accuracy compared to the other models, making it effective in detecting and understanding attack patterns hidden in the CICIDS2017 dataset. Leveraging the power of ensemble learning, Random Forest exhibited improved accuracy and robustness by combining multiple weak prediction decision trees. However, Gradient Boosting, with its iterative optimization approach, did not show significantly enhanced predictive performance, resulting in lower accuracy and higher error rates compared to the Decision Tree model.

In contrast, XGBoost, a state-of-the-art gradient boosting framework, showcased exceptional performance with high accuracy. The choice of the most suitable model depends on the specific requirements of the intrusion detection task, considering factors such as interpretability, accuracy, and computational resources. This benchmarking study provides valuable insights for practitioners in selecting the optimal decision tree family model for their specific needs.

Further research can focus on exploring advanced optimization techniques and evaluating these models on diverse datasets to validate their performance across different scenarios. Overall, this study enhances the understanding of decision tree family models in the context of intrusion detection, contributing to the development of more effective and reliable cybersecurity systems.

References

1. Mazini, M., Shirazi, B., Mahdavi, I.: Anomaly network-based intrusion detection system using a reliable hybrid artificial bee colony and AdaBoost algorithms. J. King Saud Univ. – Comput. Inf. Sci. **31**(4), 541–553 (2019)
2. Khraisat, A., Alazab, A.: A critical review of intrusion detection systems in the internet of things: techniques, deployment strategy, validation strategy, attacks, public datasets and challenges. Cybersecurity **4**(1), 1–27 (2021). https://doi.org/10.1186/s42400-021-00077-7
3. Mhawi, D.N., Aldallal, A., Hassan, S.: Advanced feature-selection-based hybrid ensemble learning algorithms for network intrusion detection systems. Symmetry **14**, 1461 (2022)
4. Maseer, Z.K., Yusof, R., Bahaman, N., Mostafa, S.A., Foozy, C.F.M.: Benchmarking of machine learning for anomaly based intrusion detection systems in the CICIDS2017 dataset. IEEE Access **9**, 22351–22370 (2021)
5. Yulianto, A., Sukarno, P., Suwastika, N.A.: Improving AdaBoost-based intrusion detection system (IDS) performance on CIC IDS 2017 dataset. J. Phys. Conf. Series **1192**, 012018 (2019)
6. Kurniabudi, D.S., Darmawijoyo, M.Y., Idris, B., Bamhdi, A.M., Budiarto, R.: CICIDS-2017 dataset feature analysis with information gain for anomaly detection. IEEE Access **8**, 132911–132921 (2020). https://doi.org/10.1109/ACCESS.2020.3009843
7. Sharafaldin, I., Habibi Lashkari, A., Ghorbani, A.A.: A detailed analysis of the CICIDS2017 data set. In: Mori, P., Furnell, S., Camp, O. (eds.) ICISSP 2018. CCIS, vol. 977, pp. 172–188. Springer, Cham (2019). https://doi.org/10.1007/978-3-030-25109-3_9
8. Reis, B., Maia, E., Praça, I.: Selection and performance analysis of CICIDS2017 features importance. In: Benzekri, A., Barbeau, M., Gong, G., Laborde, R., Garcia-Alfaro, J. (eds.) FPS 2019. LNCS, vol. 12056, pp. 56–71. Springer, Cham (2020). https://doi.org/10.1007/978-3-030-45371-8_4
9. Sharafaldin, I., Lashkari, A.H., Ghorbani, A.: Toward generating a new intrusion detection dataset and intrusion traffic characterization. In: Proceedings of the 4th International Conference on Information Systems Security and Privacy – ICISSP, ISBN 978-989-758-282-0, ISSN 2184-4356, pp. 108–116. SciTePress (2018)

10. Aksu, D., Üstebay, S., Aydin, M.A., Atmaca, T.: Intrusion detection with comparative analysis of supervised learning techniques and fisher score feature selection algorithm. In: Czachórski, T., Gelenbe, E., Grochla, K., Lent, R. (eds.) Computer and Information Sciences. ISCIS 2018. Communications in Computer and Information Science, vol. 935. Springer, Cham. https://doi.org/10.1007/978-3-030-00840-6_16

11. Tsai, J.J.P., Yu, Z.: Intrusion Detection: A Machine Learning Approach. Imperial College Press, GBR (2011)

12. Bhavani, T.T., Rao, M.K., Reddy, A.M.: Network intrusion detection system using random forest and decision tree machine learning techniques. In: Luhach, A.K., Kosa, J.A., Poonia, R.C., Gao, X.-Z., Singh, D. (eds.) First International Conference on Sustainable Technologies for Computational Intelligence. AISC, vol. 1045, pp. 637–643. Springer, Singapore (2020). https://doi.org/10.1007/978-981-15-0029-9_50

13. Brochu, E., Cora, V.M., de Freitas, N.: A Tutorial on Bayesian Optimization of Expensive Cost Functions, with Application to Active User Modeling and Hierarchical Reinforcement Learning. ArXiv abs/1012.2599 (2010)

14. Galuzzi, B.G., Giordani, I., Candelieri, A., Perego, R., Archetti, F.: Hyperparameter optimization for recommender systems through Bayesian optimization. CMS **17**(4), 495–515 (2020). https://doi.org/10.1007/s10287-020-00376-3

15. Masum, M., et al.: Bayesian hyperparameter optimization for deep neural network-based network intrusion detection. In: 2021 IEEE International Conference on Big Data (Big Data), Orlando, FL, USA, pp. 5413–5419 (2021)

16. Hodo, E., Bellekens, X., Hamilton, A., Tachtatzis, C., Atkinson, R.: Shallow and Deep Networks Intrusion Detection System: A Taxonomy and Survey (2017)

17. Axelsson, S.: Intrusion Detection Systems: A Survey and Taxonomy (2000)

Author Index

© The Editor(s) (if applicable) and The Author(s), under exclusive license
to Springer Nature Switzerland AG 2023
R. El Ayachi et al. (Eds.): CBI 2023, LNBIP 484, p. 221, 2023.
https://doi.org/10.1007/978-3-031-37872-0

Printed in the United States
by Baker & Taylor Publisher Services